Cambridge Elements ☰

Elements in the Philosophy of Science
edited by
Jacob Stegenga
University of Cambridge

BIG DATA

Wolfgang Pietsch
Technical University of Munich

CAMBRIDGE
UNIVERSITY PRESS

University Printing House, Cambridge CB2 8BS, United Kingdom

One Liberty Plaza, 20th Floor, New York, NY 10006, USA

477 Williamstown Road, Port Melbourne, VIC 3207, Australia

314–321, 3rd Floor, Plot 3, Splendor Forum, Jasola District Centre,
New Delhi – 110025, India

79 Anson Road, #06–04/06, Singapore 079906

Cambridge University Press is part of the University of Cambridge.

It furthers the University's mission by disseminating knowledge in the pursuit of
education, learning, and research at the highest international levels of excellence.

www.cambridge.org
Information on this title: www.cambridge.org/9781108706698
DOI: 10.1017/9781108588676

First published 2021

A catalogue record for this publication is available from the British Library.

ISBN 978-1-108-70669-8 Paperback
ISSN 2517-7273 (online)
ISSN 2517-7265 (print)

Big Data

Elements in the Philosophy of Science

DOI: 10.1017/9781108588676
First published online: January 2021

Wolfgang Pietsch
Technical University of Munich

Author for correspondence: Wolfgang Pietsch, wolfgang.pietsch@tum.de

Abstract: Big data and methods for analyzing large data sets such as machine learning have in recent times deeply transformed scientific practice in many fields. However, an epistemological study of these novel tools is still largely lacking. After a conceptual analysis of the notion of data and a brief introduction to the methodological dichotomy between inductivism and hypothetico-deductivism, several controversial theses regarding big data approaches are discussed. These include whether correlation replaces causation, whether the end of theory is in sight, and whether big data approaches constitute entirely novel scientific methodology. In this Element, I defend an inductivist view of big data research and argue that the type of induction employed by the most successful big data algorithms is variational induction in the tradition of Mill's methods. Based on this insight, the abovementioned epistemological issues can be systematically addressed.

Keywords: big data, machine learning, induction, causation, theory-ladenness

ISBNs: 9781108706698 (PB), 9781108588676 (OC)
ISSNs: 2517-7273 (online), 2517-7265 (print)

Contents

1 Introduction

In a provocative article in the technology magazine *WIRED*, then editor-in-chief Chris Anderson notoriously claimed that "[t]he new availability of huge amounts of data, along with the statistical tools to crunch these numbers, offers a whole new way of understanding the world. Correlation supersedes causation, and science can advance even without coherent models, unified theories, or really any mechanistic explanation at all" (Anderson 2008). Already in this brief quote, a variety of topics are raised that have long been controversially debated in epistemology and the philosophy of science; for example, the role of causation in scientific research or the importance of models and theories for understanding the world. Anderson picked up his views at least partly from discussions in the machine learning community. As outlandish as his claims may sound to many scientists and professional philosophers, Anderson's article nevertheless hit a nerve and marked the beginning of an ongoing debate about the impact of modern information and communication technology on the scientific method.

Many reactions have been highly critical. Few have deemed Anderson's theses worthy of academic scrutiny, and thus the debate was mainly confined to informal discussions, newspaper articles, or blog posts. For example, in reply to Anderson's depiction of big data approaches as being theory-free, the allegedly indispensable role of hypotheses for the scientific method has often been stressed. Sean Carroll, a theoretical physicist at Caltech and well-known author of popular science books, writes that "[h]ypotheses aren't simply useful tools in some potentially-outmoded vision of science; they are the whole point. Theory is understanding, and understanding our world is what science is all about."[1] Werner Callebaut, a philosopher of biology, has responded in a similar manner: "With the theories and models and the scientific method in the bathwater, the baby has gone as well. Anderson's argument is so obviously flawed that I wouldn't have referred to it at all hadn't it become so influential" (Callebaut 2012, 74). Unfortunately, many responses have been almost as simplistic as Anderson's theses themselves, and they are often based on substantial misunderstandings and misconceptions of scientific methodology.

In this Element, I will defend a middle ground arguing that Anderson's theses, although much overblown, constitute useful starting points for an epistemological discussion of big data approaches. Maybe the most convincing argument to the effect that big data changes science in fundamental ways consists in a number of impressive success stories, where large data sets have

---

[1] See www.edge.org/discourse/the_end_of_theory.html for a variety of responses to Anderson by scientists and science writers.

played a crucial role in solving difficult scientific problems. These problems almost all concern the largely automated prediction of phenomena based on circumstances of these phenomena. There may still be comparatively few of such success stories, but they should be considered as proof of concept that an inductivist big data approach can be fruitfully applied to certain types of research tasks rather than staking out the entire application spectrum. As will be seen, some of the criticism raised against big data approaches is already invalidated by these examples from scientific practice.

For instance, large data sets have long been used in linguistics for diverse tasks such as speech recognition, machine translation, spelling correction, or author identification of a text. As any user of these tools can confirm, the quality of results has dramatically improved over the last two decades. This improvement can in part be attributed to better and bigger data sets being available for analysis, but also to more sophisticated machine learning algorithms, for example the renaissance of neural networks in recent years, which proved to be a breakthrough technology for many fields. The modeling often does not require any linguistic theory at all. As an example, machine translation in a somewhat primitive but still moderately successful form is based merely on frequency data of so-called n-grams in large text corpora, that is, word sequences of length n (e.g., Jelinek 2009; Norvig 2009).

Big data and machine learning have also been fruitfully applied in the medical and health sciences. For example, malignant skin cancer has been identified based on images of skin lesions using deep neural networks with an accuracy comparable to that of human experts. The classification was based on large data sets of correctly classified images without substantial further biological or medical background knowledge (Esteva et al. 2017). Image recognition technology has also been used for the analysis and prognosis of lung cancer based on histopathology slides (Yu et al. 2016). In fact, these are just two examples of a wide variety of medical image recognition tasks, to which machine learning algorithms are applicable.

Since much of the data that is available and collected on the Internet and other computer networks is at least partially of a social nature, it is hardly surprising that big data approaches have been tried with various degrees of success in the social sciences. The big tech firms like Facebook, Google, and Amazon are constantly conducting experiments to determine how to best influence or manipulate users (Kohavi et al. 2020; Luca & Bazerman 2020). Face recognition is another example of a task with relevance for social science problems, for which big data approaches have at times superseded human capabilities (Ghani & Schierholz 2017, 149). It is also possible to identify humans on the basis of their walking or gait characteristics (Wan et al. 2019) or of their heartbeat patterns visible from afar and through clothing in infrared spectrum (Hambling 2019).

The examples described above illustrate well the type of problems for which big data analytics holds the largest promises, essentially situations in which specific variables are to be predicted based on other variables characterizing a phenomenon. Thus, big data analytics understood as the application of machine learning algorithms to large data sets is not likely to lead to any of the grand theories for which scientists have usually been admired such as the theory of relativity or the theory of evolution. Big data analytics is concerned with more basic tasks in scientific methodology; for example, with classification or with the identification of phenomenological laws, which may hold only locally in specific contexts in contrast to the fundamental laws of grand theoretical schemes.

While these examples leave little doubt that big data approaches are occasionally working, there is a large grey area, for which predictive success is much murkier and more difficult to pinpoint. Many uses of big data in business or government are of this murky type.[2] Sometimes, the success of big data approaches is ephemeral. A well-known example in this respect is Google Flu Trends, which has been portrayed as a "parable" of the "traps in big data analysis" (Lazer et al. 2014). Google Flu Trends had been set up as a predictive engine that could predict flu patterns based on the terms that people enter in Google's search engine. It turned out that this worked well for a few years, but eventually, for reasons that are not entirely clear, predictions started to deviate substantially from the data provided by doctors. It has often been taken as a fundamental lesson from this episode that traditional principles of data preparation and statistical analysis have to be adhered to also in the age of big data. But while it is interesting to ponder what went wrong in this specific example, it certainly cannot invalidate the big data approach as a whole. For instance, it is highly implausible that machine translation or algorithmic skin cancer recognition will suddenly cease to give viable results in an analogous manner.

In the following sections, I will critically assess central epistemological theses with respect to big data,[3] in particular whether correlation replaces causation in big data research (Section 5), whether big data analysis can

---

[2] Compare Robert Northcott's argument starting from case studies mainly in economics and the social sciences that the prospects for improved predictions based on big data are rather limited (2019).

[3] See, for example, Anderson 2008; Kitchin 2014; Mayer-Schönberger & Cukier 2013. For an influential and insightful early epistemological analysis of related issues, see Gillies (1996).

dispense with background theory (Section 6), and whether big data analysis constitutes an entirely new type of scientific research (Section 7).

Section 2 lays some of the conceptual foundations, in particular by defining the notion of 'data' and by discussing, based on this definition, the epistemological role of data in processes of learning about the world.

Section 3 examines to what extent big data analytics constitutes an inductivist approach inferring general laws directly from data or whether it embodies hypothetico-deductivism, which is at present widely held to be the only feasible scientific method. As I will argue, machine learning approaches applied to large data sets are clearly inductivist.

Section 4 examines in more detail the type of induction that underlies the most successful machine learning algorithms used for analyzing big data, in particular, decision trees and neural networks. I argue that these algorithms all implement a variational type of induction closely related to John Stuart Mill's canons of induction. This central result will provide the basis for subsequent discussions of the role of causation and the role of theory in big data approaches.

Section 5 assesses whether big data approaches exclusively rely on the search for correlations in large data sets and abandon any quest for causality. Starting from the results of Section 4, I show that various machine learning algorithms yield causal information because variational induction typically identifies causal relationships. Furthermore, I argue that predictive success is possible only if big data algorithms manage to at least approximate causal relationships. Thus, contrary to popular opinion, causation is actually a crucial concept to understand the success of data-scientific practices.

Section 6 looks at the role of theory, asking which kinds of modeling assumptions are required in big data approaches and which can be dispensed with. This analysis will also build on the results of Section 4 in that the answer is in part provided by determining the theoretical assumptions that have to be presupposed in variational induction.

Finally, Section 7 examines how novel big data approaches are in terms of scientific methodology by drawing a parallel to exploratory experimentation as a long-established research practice.

2 Defining Big Data

2.1 Why Definitions Matter

Looking back into the history of science, influential scientists, in particular those working on foundational issues in various fields, have repeatedly acknowledged the significance of precisely defined and contextually adequate scientific language. A well-known example in this respect is Antoine Lavoisier, who is widely considered to be one of the founders of modern chemistry. Lavoisier was the central figure of the so-called chemical revolution toward the end of the eighteenth century, during which basic tenets of modern chemistry such as the law of conservation of mass or the oxygen theory of combustion were established.

In the introduction to his magnum opus, *Elementary Treatise of Chemistry*, Lavoisier recounts the process by which he arrived at his major scientific results: "while I thought myself employed only in forming a Nomenclature, and while I proposed to myself nothing more than to improve the chemical language, my work transformed itself by degrees, without my being able to prevent it, into a treatise upon the Elements of Chemistry" (Lavoisier 1789, xiv). He concluded that "we cannot improve the language of any science without at the same time improving the science itself; neither can we, on the other hand, improve a science, without improving the language or nomenclature which belongs to it" (Lavoisier 1789, xv).

By analogy, the epistemological questions regarding big data approaches that were raised in the introduction can only be tackled based on a precise and coherent conceptual framework. Indeed, much of the confusion surrounding the foundations and possibilities of big data approaches stems from imprecise language or even outright conceptual misunderstandings. Accordingly, the answers given in the course of this Element will mostly flow from carefully developed definitions and explications of fundamental methodological terms with relevance for big data like induction, causation, or the notion of data itself. I will turn now to the latter concept.

2.2 Defining Data

The notion of 'data' has thus far received comparably little attention in the epistemological literature. In the following sections, various definitions of data that have been proposed will be critically assessed in view of the role that these definitions could play for an epistemological analysis of data science. Not least, data needs to be clearly defined in order to distinguish the term from related concepts such as information, facts, or phenomena and to clarify the exact role that data can play in practices like data science for learning about the world.

2.2.1 Computational Definitions

Computational definitions of data start from the central role data has played for the rise of information and communication technology in past decades. These definitions try to capture a commonsense notion of 'data' based on current usage of the term in many of the empirical sciences and in computer science. They generally explicate data by referring either to the medium on which data is stored or to the format in which it is stored.

Computational definitions draw on a long tradition, within which the notion of 'data' has been defined as *marks* or *traces*, that is, as uninterpreted inscriptions that may, for example, be the outcomes of an experiment or manipulations of those outcomes (e.g., Hacking 1992, 43–4, 48–50; Rheinberger 2011; cf. also Leonelli 2016, 75–6). Accordingly, in Ian Hacking's account, the making and taking of data and the interpreting of data can be distinguished as subsequent steps. Hacking develops his notion of data with a keen eye on scientific practice, emphasizing the role of statistics in all kinds of data processing such as data assessment, reduction, or analysis (Hacking 1992, 48–50). Similarly, a central aspect of the modern notion of data as used in many of the empirical sciences is that data can be aggregated and further analyzed by computational systems. This has inspired the terminology of 'computational definitions' as used by various authors such as Luciano Floridi (2008, 2011)[4] or Aidan Lyon (2016).

Variants of the computational definition understand data as *anything that is stored on a hard drive* or as *anything that is stored in a binary manner* (e.g., Floridi 2008). Such definitions may be adequate for the contemporary data scientist, since most of today's data is stored on hard drives in a digital format. However, it is plausible to assume that data existed before hard drives were used and before binary representations of information became standard. For example, there is no good reason why the observations of the stars, which the astronomer Tycho Brahe meticulously recorded in the sixteenth century, should not be classified as data.

A computational definition of data, which applies equally to data predating the modern digital age, could read:

> Data is anything that is stored in a symbolic form on a medium.

Such a definition captures an important insight regarding the nature of data, namely that data should be distinguished from the thing itself, which the data is about (cf. Hacking 1992, 44). For example, music data certainly is not music, but is about music. Equally, scientific data must not be conflated with the phenomenon to which it refers. Data always *represent* an entity or a process, but are in general not identical with that entity or process. The notion of a 'symbolic form' captures this representational nature of data quite well

---

[4] Floridi, who is generally regarded as a founding figure of the emerging field of philosophy of information, has developed the so-called diaphoric interpretation of data, according to which data are ultimately reducible to a lack of uniformity. This interpretation should be understood as underpinning his notion of 'semantic information' being "well-formed, meaningful and truthful data" (2011, 104; see also Adriaans 2019; Floridi 2019).

since symbols are generally considered to be signs that, if correctly interpreted, refer to something else. For comparison, Ian Hacking chose the term 'mark' in part because its meaning is closely related to concepts like 'sign' or 'symbol' (Hacking 1992, 44). Symbols may, for example, be numbers, letters, graphs, or pixels. A 'medium' can be any material carrier of such symbolic forms, i.e., any material entity into which symbolic forms may be inscribed. Media can be hard drives, books, or photographic paper.

However, for answering the epistemological questions that are of interest in the present Element, computational definitions are too broad. If anything that is stored in symbolic form on a medium constitutes data, such data could be texts, images, videos, or music clips without any scientific or epistemological significance whatsoever. Thus, what is missing is an epistemological link to how data can be used for learning about the world. While such a link does not have to be part of the definition of data, it should at least follow naturally from the definition. Thus, the main criticism of computational definitions from an epistemological perspective is that these provide no hint of the role that data can play in the scientific method.

To be clear, the computational definitions given above are not wrong and they may be useful for data practitioners who are concerned with all kinds of ways of dealing with data, e.g., data storage, compression, or encryption. But such computational definitions are not specific enough for the task of this Element, namely to gauge the *epistemological* significance of data and in particular of big data.

2.2.2 The Representational Definition

Representational accounts of 'data' focus on the widely accepted idea that data always represent certain facts or phenomena (for a thorough discussion see Leonelli 2016, Sect. 3.1). For example, in a recent survey paper, Aidan Lyon proposes the following variant of a representational definition:

> Data are representations of purported facts. (2016, 752)

Lyon introduces a distinction between 'fact data' and 'fictional data', wherein the above definition is supposed to capture the notion of fact data. Fictional data represent intended commands, as is, for example, the case for some music data, video data, image data, or code data.

Lyon emphasizes that fact data are different from the facts themselves, because fact data can be "stored, compressed, destroyed, encrypted, and hidden [while] facts are not the sorts of things that can have all of these things done to them" (2016, 740). Furthermore, according to Lyon, fact data can be false or falsified, while facts themselves cannot. Lyon illustrates this point with the well-known hypothesis proposed by the statistician Ronald Fisher that Gregor Mendel's data on inheritance may have been fabricated by one of Mendel's assistants, who knew a little too well what results were to be expected.

Indeed, the story of Mendel and his assistant can be used to illustrate two central conceptual issues arising with respect to the notion of data: first, as already mentioned, whether data themselves can be false, and second, whether data needs interpretation. These questions are to some extent related. In particular, it seems plausible to assume that data can be false or fake only in light of a specific interpretation of the data.

In order to allow for false data, Lyon relates data not with the facts but with *purported* facts, that is, with what one thinks are the facts (2016, 740). However, by making this definitional choice, Lyon implies that all data are to some extent interpreted. After all, purporting facts, which are allegedly represented by the data, constitutes a crucial part of any interpretation. Conversely, Lyon's solution to the problem of false data is not available if data *per se* are taken to be uninterpreted as, for example, in Hacking's account.

If an interpretational step is not integrated into the definition of data, one can take a different approach by assuming that data themselves cannot be false, but can merely be falsely *interpreted*. According to this view, truth and usefulness depend not on the data themselves but on the specific interpretation. In the example of Mendel's assistant, the data were deliberately fabricated to be falsely interpreted as representing certain facts about the inheritance of traits in the crossbreeding of peas, while in reality they do not represent those facts. However, the same data may be very valuable when interpreted as indicating the state of mind or the expectations of Mendel's assistant.

From this perspective, it appears sensible to define data as representations merely of facts rather than of purported facts. Still, when looking at scientific practice, there is little reason to deny the core idea of the representational definition that representation is a central function of data. This representational nature results because the data are in some objective way related to the facts they represent, e.g., causally related.

2.2.3 The Relational Definition

In an acclaimed treatise (2016) on the role of data in modern biology, which has been awarded the prestigious Lakatos Award in philosophy of science, Sabina Leonelli rejects representational definitions and endorses instead what she terms a relational definition of data. The relational definition has two essential features:

any object can be considered as a datum as long as (1) it is treated as (at least potential) evidence for one or more claims about the world, and (2) it is possible to circulate it among individuals/groups. (Leonelli 2019, 16; see also Leonelli 2016, 78)

The first part of the definition captures the idea that the evidential value of data is externally ascribed and depends on the way data is handled. The content of data thus is determined by "their function within specific processes of inquiry" (2019, 16). In that sense, data is 'relational' by always referring to a certain context of usage – hence the name for her approach. By defining data in terms of their evidential function, Leonelli draws an explicit connection to the epistemological question, what role data can play in learning about the world. This conceptual link constitutes a clear advantage of her approach, at least when it comes to addressing epistemological issues related to data science. The first part of Leonelli's definition reflects a key lesson from an influential analysis of the data-phenomena distinction by James Bogen and James Woodward (1988). For comparison, Woodward defines 'data' as "public records produced by measurement and experiment that serve as evidence for the existence or features of phenomena" (2011, 166).

The second part of Leonelli's definition takes up a central theme of her analysis of data-centric biology according to which the "real source of innovation in current [data-centric] biology is the attention paid to data handling and dissemination practices and the ways in which such practices mirror economic and political modes of interaction and decision making, rather than the emergence of big data and associated methods per se" (2016, 1). In her book, Leonelli describes in much detail what she fittingly calls "data journeys," that is, how data travel, how they are collected, packaged, processed, and analyzed. There, she focuses on "the material, social, and institutional circumstances by which data are packaged and transported across research situations, so as to function as evidence for a variety of knowledge claims" (2016, 5).

However, this second criterion of Leonelli's definition turns out to be problematic as a basis for addressing epistemological issues related to big data and machine learning. By examining how data travel between different scientists and scientific communities, Leonelli has identified an important topic that has had an enormous impact on the way science is conducted in the digital age. However, definitions generally list only necessary conditions of the entity to be defined, i.e., the definiendum. At least for the epistemological issues discussed in the following sections, the communicability of data between individuals and groups should not be taken as a necessary criterion for data. Otherwise, the data of a lone but maybe genial scientist, who wants to keep her data to herself and develops a cabbalistic encryption for her data that is incomprehensible to everyone except herself, would not count as data. Furthermore and even more troublesome, the second part of Leonelli's definition renders certain types of data-based automated science impossible *by definition*.

By requiring that data necessarily is communicable between people, between individuals and groups, Leonelli excludes from the outset the possibility of a machine science that operates largely independently of human intervention, a machine science that can collect and analyze quantities and formats of data that may not be accessible to the human mind at all. This is one of the fundamental questions examined in this Element: to what extent is such an automated machine science feasible? Therefore, the answer to this question should not be predetermined by choosing overly narrow definitions for certain fundamental terms. Thus, the second part of Leonelli's definition may be an adequate starting point for the social and institutional studies of the impact of big data, in which she is interested. However, it is inadequate for the epistemological analysis of this Element.

Having introduced the relational definition, let us briefly address whether Leonelli's critique of representational definitions succeeds. For Leonelli, representational notions of 'data' are characterized by the idea that data always have "some sort of representational content, in the sense of instantiating some of the properties of a given target of investigation in ways that are mind-independent" (Leonelli 2019, 4). By contrast, Leonelli argues that, in the absence of a specific modeling context, data do not have any such content. For someone who disagrees with Leonelli on the issue of the representational nature of data, Leonelli thus raises the challenge to clarify what the objective mind-independent content of data could be.

One possibility is that this content is determined by causal relationships between data and facts. Such causal relationships plausibly exist if the data record experimental results or observations. For example, the pixels of a photograph have reliable causal relationships to the facts that are depicted in the photograph. These causal relationships result from the radiation emitted by the phenomenon of investigation and received by the photographic sensor. When the representation changes – for example, when the same photograph is described by words in a textbook – an additional transformation is applied to the data. Such transformations are typically based on established definitions and conventions of ordinary or scientific language. That the representational content of the data is essentially determined by causal relations with the represented facts seems to be quite universal for the scientific data we are interested in. As the discussion of causation in Section 5 will establish, causal relationships can be objective and largely independent of specific contexts of inquiry. This casts doubt on Leonelli's claim that data cannot have context-independent content and thus on her main argument against representational definitions.

2.2.4 Summarizing

As has now repeatedly been emphasized, fundamental notions like data should not be defined independently of the respective context in which they are to be used. In this Element, we are interested in answering epistemological questions concerning big data and data science. Therefore, a corresponding notion of data has to be integrated into a broader picture of how we acquire knowledge about the world. Following is a list of features that seem adequate for such a notion of data, all of which are either self-explanatory or have been motivated in the above discussion of this section (cf. Fig. 1):

(i) Data are not the facts themselves, but rather traces or marks of these facts.
(ii) The traces or marks require a physical medium in which they are inscribed.
(iii) In contrast to the facts themselves, the traces or marks must have a certain persistence on the medium so that they can be aggregated and analyzed.
(iv) The facts are about a phenomenon of interest, which is chosen by an observer from a number of different possibilities.
(v) The facts are singular facts referring to a specific instantiation of the phenomenon of interest, for example, an individual event or object (token).
(vi) The traces or marks must have some meaningful relationships with the facts, typically causal and/or definitional.

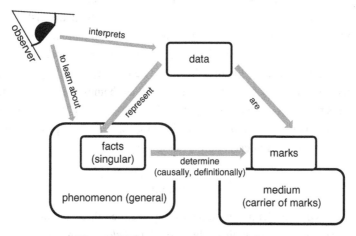

Figure 1 The proposed notion of data.

(vii)   Because of these relationships, data represent the facts at least partially.
(viii) Because data are different from the facts, they need to be interpreted in order to learn about the phenomenon of interest from the data.
(ix)   It must be possible to render data in a form such that an appropriate scientific method can be applied to the data.
(x)   By means of such a scientific method, data can serve as evidence for phenomena, which are general (types).

If one had to summarize the most important aspects about data in a single definition, such a definition might read as follows:

> Data are marks on a physical medium (such as a piece of paper, a hard drive, or an electromagnetic wave) that are meaningfully (e.g., causally or definitionally) related with certain singular facts belonging to a phenomenon of interest. If the data are correctly interpreted in terms of the relationship that they have with those facts, then the data constitute evidence for those facts and thus the phenomenon of interest.

This definition takes into account the various arguments presented in the preceding discussion of the different interpretations of data. In particular, it accounts for the distinction between data and facts and specifies the relationship between data and facts as typically being causal or definitional. With respect to the problem of false data, it explains how the data themselves cannot be false but can be wrongly interpreted by someone unaware of the exact relationship between data and facts. Regarding the evidential role of data, it situates the role of data within an epistemological procedure of learning about the world. By clarifying that data represent singular facts, it accounts for the widespread intuition that data are very close to immediate, unfiltered experiences about the world. Letting data represent singular facts also allows one to address a problem raised by Leonelli (2019), how to conceptually distinguish between data and models. According to the above definition, data represent tokens, i.e., individual instances, while models represent types of phenomena, which generalize to a certain extent. Finally, the above definition allows for all kinds of epistemic agents to process data, including human beings, but also machines. This seems to be a good starting point for addressing epistemological questions related to big data; for example, whether a machine science processing data largely without human intervention is possible at all.

2.3 When Is Data Big?

Big data is typically defined in terms of the so-called three V's as data that is large in *Volume*, has high throughput or data *Velocity*, and is provided in a *Variety* of different formats (Laney 2001). Further dimensions or V's are occasionally added, for example, veracity, value, variability, or visualization (Vo & Silva 2017, 125). Some of these definitions determine a threshold, however vague, above which data sets become big. For example, big data has been described as being "too big to fit onto the analyst's computer" (2017, 125). Wikipedia at one point characterized big data as being "so large and complex that traditional data processing applications are inadequate" (2017, 125).

The above definitions of 'bigness' all refer to the computational challenges that arise with large data sets. Accordingly, these are adequate definitions from a computational point of view, since they point to technical problems resulting when data cannot anymore be handled by means of conventional tools. For example, novel database management systems are needed to cope with extremely large and unstructured data sets, indicating interesting paradigm changes in the handling of data (e.g., Foster & Heus 2017). Also, powerful parallel computing systems such as Apache Hadoop are increasingly employed, because many large data sets cannot be processed by a single computer or processor alone (e.g., Vo & Silva 2017). However, none of these developments has obvious ramifications for the scientific method.

Other scholars have an interest in the wider implications of big data beyond the technical issues mentioned earlier. Accordingly, they frame 'big data' in very broad terms as a complex interplay of social and technological practices. Thus, the term 'big data' ceases to refer to large data sets as such, but designates broader socio-technical phenomena, which are somehow related to big data. An example in this respect is the following influential definition proposed by danah boyd and Kate Crawford, both of Microsoft Research, in an article that is mainly critical of big data:

> We define Big Data as a cultural, technological, and scholarly phenomenon that rests on the interplay of: (1) Technology: maximizing computation power and algorithmic accuracy to gather, analyze, link, and compare large data sets. (2) Analysis: drawing on large data sets to identify patterns in order to make economic, social, technical, and legal claims. (3) Mythology: the widespread belief that large data sets offer a higher form of intelligence and knowledge that can generate insights that were previously impossible, with the aura of truth, objectivity, and accuracy. (2012, 663)

There is certainly merit to identifying the impact of big data in culture and society beyond its mere scientific and technological significance. While the definition just presented is certainly useful for sociological analyses, it is inadequate to address the epistemological questions that I aim to study in this Element. Importantly, boyd and Crawford's definition fails to identify a clear-cut criterion for when data should be considered 'big'.

There is, however, one definition of 'bigness' that provides a promising starting point for addressing epistemological questions. According to this definition, data are considered as 'big' once they cover a whole domain of interest, that is, once they capture "N = all" (e.g., Mayer-Schönberger & Cukier 2013, 197; Kitchin 2014, 1).[5] According to this idea, big data records every individual in the examined population,[6] not only a representative sample.[7] If still somewhat imprecise, this approach does have epistemological implications. Maybe most importantly, such a definition could explain why big data leads to changes in modeling. If it is possible to look up every individual in a database, sophisticated theoretical modeling may become dispensable. After all, the information is already in the data.

While these ideas have some appeal, there are various problems with a definition of 'bigness' in terms of "N = all." Most obviously, it is generally impossible to include every individual in a data set. For example, some individuals of a population may become available only in the future or may have been available only in the past. A further problem is that N could easily become unrealistically large so that even big data technologies will be unable to deal with such amounts of data.

For these reasons prediction or classification based merely on a lookup table, in which everything is documented, is too simplistic. Instead, the specific inductive methods on which big data analysis relies need to be taken into account, including in particular machine learning approaches. Accordingly, a data set may be defined as 'big', if it comprises sufficient data points so that inductive inferences based on the data set yield reliable predictions. Thus defined, big data indeed covers a whole domain; however, not in the literal sense of a lookup table, but rather with reference to specific inductive methods.

A further, more cosmetic problem for the "N = all" approach is that, in the case of simple domains, small data sets will often suffice for predictions. This has the somewhat paradoxical consequence that, if 'big data' is defined in terms of "N = all," small data fully describing simple domains would already qualify as "big" data. To prevent this counterintuitive implication, only domains

---

[5] The "N = all" idea is reminiscent of the microarray paradigm introduced by Napoletani et al.: "we shall term *microarray paradigm* the modus operandi [. . .] which we can summarize as follows: *if we collect enough and sufficiently diverse data regarding a phenomenon, we can answer most relevant questions concerning the phenomenon itself*" (2011, italics in the original; cf. also Panza et al. 2011).

[6] Here, 'population' should be understood in a general sense as referring to a set of similar objects or events.

[7] Plantin and Russo (2016) discuss whether analyzing all data instead of inferring from samples allows for a return to determinism.

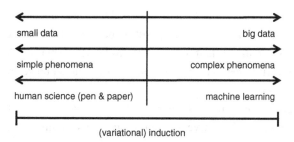

Figure 2 Big data versus small data in scientific research.

comprising sufficiently complex phenomena should be included in the definition of 'big data.'[8] Then, the corresponding data sets are necessarily large, often larger than can be handled by an individual human being (see Fig. 2).

The refinements of the "N = all" approach just presented are summarized in the following definition:

> A data set is 'big' if it is large enough to allow for reliable predictions based on inductive methods in a domain comprising complex phenomena.

For complex phenomena, small data would not enable reliable predictions. Note further that according to this definition some large data sets do not count as big data, namely if they do not allow for reliable predictions or if they represent only simple phenomena. This definitional choice sets the focus on epistemological issues and clarifies that the term 'big data' is appropriate only if the data are scientifically useful. Finally, the definition emphasizes that the epistemological implications of big data can be determined only in relation to specific inductive methodology, a topic that will be addressed in Sections 3 and 4.

3 Inductivism

3.1 The Debate on Inductivism in Data Science

Some big data advocates have argued that data, if only they are big enough, can speak for themselves (Anderson 2008). The underlying conception is that data-scientific practices start with the data and infer general laws or models directly from the data with little or no additional theoretical assumptions. Epistemologically speaking, this would be an inductivist approach. *Inductivism*, of course, is the very idea that science starts with the facts and derives general laws and models from those facts by means of inductive inferences. The methodological antithesis to inductivism is *hypothetico-deductivism*, according to which science

---

[8] Complexity, in the present context, should be understood broadly in terms of a complicated structure (cf. Kuhlmann 2011).

begins with hypotheses about general laws and then uses deductive inferences to derive statements of facts, which are subsequently tested by experiment and observation. A popular variant of hypothetico-deductivism is *falsificationism*, emphasizing Karl Popper's idea that hypotheses can never be verified but only be falsified (1935).

In this section, I examine whether big data approaches constitute inductivist or hypothetico-deductive methodology. This is important not least because inductive methods can be automated far more easily compared with a hypothetico-deductive or falsificationist approach. After all, inductive methods are often quite simple and can be carried out quasi-algorithmically, while the formulation of hypotheses requires creativity and intuition, which cannot be easily algorithmized (e.g., Williamson 2009, 80). This point was already acknowledged in the seventeenth century by Francis Bacon, often considered the father of inductivism, who noted that the inductive business could "be done as if by machinery" (Bacon 1620, 256; cited in Gillies 1996, 2).

However, inductivism has been largely rejected in twentieth-century epistemology and philosophy of science based on arguments by Pierre Duhem (1906), Karl Popper (1935), and others. In many empirical sciences, some version of hypothetico-deductivism is often considered the only viable scientific method.[9] When big data advocates argue for inductivism, it is thus often objected that they ignore even the most basic insights of modern philosophy of science (see, e.g., the quotes by Carroll and Callebaut in the introduction).[10]

Such an anti-inductivist stance is put forward in an article by the information scientist Martin Frické, who summarizes the main argument against inductivism in data science as follows:

> An immediate concern about the present enchantment with data-driven Big Data is just that it might be inductivism, the hoary punching bag from the philosophy of science. [...] That inductivism is a mistaken philosophy of science is not controversial – it is received wisdom. (2014, 652–3)

---

[9] Indeed, various encyclopedias, including the venerable *Encyclopaedia Britannica* and *Wikipedia*, identify the scientific method more or less explicitly with the idea of hypothesis testing, that is, with a hypothetico-deductive approach (see the respective entries on "scientific method," consulted on September 25, 2020). The *Oxford Dictionary of Psychology* considers the hypothetico-deductive method to be the "standard research method of empirical science," which is "[a]lso called the scientific method" (Colman 2015, 358). Similarly, according to Craig Calhoun's *Dictionary of the Social Sciences*, the hypothetico-deductive method is "[o]ften referred to as the scientific method" (Calhoun 2002, 219).

[10] An exception in this respect is Jon Williamson, who notes in an epistemological discussion of machine learning that "advances in automated scientific discovery have lent plausibility to inductivist philosophy of science" (2004, 88).

Similar anti-inductivist views, if rarely expressed with the radicality of Frické, are shared by many influential scholars who have commented on the epistemology of data science.

Besides the anti-inductivist reaction, there is another more conciliatory response, which argues that the scientific approach is never purely inductive or purely hypothetico-deductive. Instead, it is always an amalgam of different methodologies. For example, Rob Kitchin, a geographer, who has written one of the first book-length treatises on big data (2014), endorses this view:

> In contrast to new forms of empiricism, data-driven science seeks to hold to the tenets of the scientific method, but is more open to using a hybrid combination of abductive, inductive and deductive approaches to advance the understanding of a phenomenon. [. . .] it forms a new mode of hypothesis generation before a deductive approach is employed. (2014, 5–6)

The idea that big data approaches provide a new mode of hypothesis generation is quite common in the literature, as it seemingly allows integrating big data and machine learning into today's dominant picture of scientific method, which is hypothetico-deductivism (cf. Mazzocchi 2015). However, at the end of the day, this idea only leads to a somewhat refined hypothetico-deductivism rather than a real synthesis of inductivism with hypothetico-deductivism. The crucial issue overlooked by the type of response exemplified with Kitchin's quote is that inductivism and hypothetico-deductivism are in many respects contradictory methodological paradigms and are not as easily reconciled as proposed by Kitchin.

Before I explain this in more detail, I should stress that the following exposition of inductivism and hypothetico-deductivism necessarily needs to simplify matters. Not all versions of inductivism and hypothetico-deductivism that have been proposed in the past strictly adhere to the below characterizations. Furthermore, some scholars deny the basic dichotomy of inductivism and hypothetico-deductivism altogether and have proposed yet other paradigms such as Bayesianism. In spite of the complexity of these debates, I believe that the discussion in Section 3.2 is to a sufficient extent historically accurate and conceptually coherent to serve as a basis for the epistemological analysis of this Element.

3.2 The Nature of Inductivism

3.2.1 Inductivism

A classic statement of an inductivist scientific methodology can be found in Jean-Marie Ampère's masterpiece, his *Mathematical Theory of Electro-*

Dynamic Phenomena Uniquely Derived from Experiments (Ampère 1826). Ampère, of course, was one of the most important contributors to the classical theory of electromagnetism. In fact, he has often been referred to as the "Newton of electromagnetism." In the introduction to his treatise, Ampère summarizes his method as follows:

> First observe the facts, while varying the conditions to the extent possible, accompany this first effort with precise measurement in order to deduce general laws based solely on experiments, and deduce therefrom, independently of all hypotheses regarding the nature of the forces which produce the phenomena, the mathematical value of these forces, that is to say, the formula which represents them, this was the path followed by Newton. This was the approach generally adopted by the scholars of France to whom physics owes the immense progress which has been made in recent times, and similarly it has guided me in all my research into electrodynamic phenomena. I have relied solely on experimentation to establish the laws of the phenomena and from them I have derived the formula which alone can represent the forces which are produced; I have not investigated the possible cause of these forces, convinced that all research of this nature must proceed from pure experimental knowledge of the laws [...] (1826, 2)

This brief statement includes all core tenets of an inductivist approach:[11] (i) Scientific laws can and should be proven by induction from the phenomena, that is, from experiment and observation ("I have solely relied on experimentation"). Here, induction can be understood in a broad sense as any systematic, nondeductive inference from particular instances to general laws or models. (ii) These laws can be considered true or at least highly probable in the sense that they adequately describe a certain range of phenomena ("the formula which alone can represent the forces"). (iii) This implies an aversion against hypotheses, which by definition are always preliminary and never proven beyond doubt ("independently of all hypotheses"). As many authors in the inductivist tradition stress, hypotheses may be and are formulated in a premature state of a scientific field, but eventually scientific knowledge should move beyond the merely hypothetical. (iv) Relatedly, background knowledge or modeling assumptions play only a minor role and are often justified by or derived from empirical data. (v) Many inductivists have assumed that scientific laws are derived by a methodology of varying the circumstances ("varying the conditions to the extent possible"). (vi) This process continuously improves the knowledge about the phenomena, at least in the long run. (vii) Finally,

---

[11] Compare Pietsch (2017). The lists in this section and Section 3.2.2 are meant as plausible reconstructions of the central characteristics of both inductivism and hypothetico-deductivism. However, given the lack of scholarly consensus regarding these methods, these lists should not be read as specifying necessary and sufficient conditions.

inductivism establishes a hierarchy of laws of increasing universality, starting with simple observation statements that are combined into low-level phenomenological laws, and then into laws of increasing generality and abstractness slowly rising until the highest level of generality.

The dominance of inductivism during considerable periods of modern science can at least in part be traced back to Francis Bacon, who by many is considered among the founders of modern empirical science. His *Novum Organum* was conceived to replace Aristotle's *Organon*, which arguably is the most influential methodological treatise of ancient and medieval science. Bacon contrasted his inductivist method of discovering truth to the approach that he considered prevalent in his days in the following manner:

> There are and can be only two ways of searching into and discovering truth. The one flies from the senses and particulars to the most general axioms, and from these principles, the truth of which it takes for settled and immoveable, proceeds to judgment and to the discovery of middle axioms. And this way is now in fashion. The other derives axioms from the senses and particulars, rising by gradual and unbroken ascent, so that it arrives at the most general axioms last of all. This is the true way, but as yet untried. (1620, I.XIX)

The first approach, described in this quote, is deductivism starting from first principles, which are assumed to be known, for example, by rational intuition. Deductivism bears some resemblance to hypothetico-deductivism, the main difference being that the latter starts with hypothetical premises rather than established first principles. Bacon's own methodology, i.e., the second approach, exhibits many of the before-mentioned characteristics of inductivism. Most importantly, Bacon insisted that science should start with the facts, with careful observation and experiment, and then slowly ascend to increasingly general laws until finally arriving at what he called the forms of the phenomena, i.e., their real nature or true causes. Furthermore, Bacon's methodology based on his famous *tables of discovery* clearly exhibits a variational element, e.g., when positive instances of a phenomenon, which are collected in the table of presence, are compared with related negative instances, which are collected in the table of deviation or absence in proximity. Indeed, Bacon considered it to be one of the major improvements of his method that he replaced naïve conceptions of induction by enumeration with more sophisticated inductive approaches (1620, 21).

For a long time, epistemologists and scientists alike referred primarily to Bacon when expounding their methodological views. Some of the most influential scientists between the seventeenth and the nineteenth centuries considered themselves to be direct heirs of Bacon's scientific methodology.

Major figures situating themselves in the empiricist and inductivist Baconian tradition were, for example, Antoine Lavoisier, Isaac Newton, William Herschel, and John Stuart Mill.

Inductivism has been under attack at least since the second half of the nineteenth century, leading to several prominent debates between inductivists and deductivists. There was a famous exchange between John Stuart Mill and William Whewell, the former siding with inductivism, the latter arguing for a hypothetico-deductivist approach, stressing the indispensable role of hypotheses in science. Several decades later, this controversy was continued between the empiricists of the Vienna Circle and Karl Popper, who took the opposing view of hypothetico-deductivism. Major arguments in these debates will be considered in Section 6, once I have introduced specific inductive methods. It is largely uncontroversial that hypothetico-deductivism emerged mostly victorious from these debates and is today often considered in the empirical sciences to be the only viable scientific methodology.[12]

3.2.2 Hypothetico-Deductivism

Let me now briefly sketch the main tenets of hypothetico-deductivism illustrated by the following quote of Albert Einstein:

> We have now assigned to reason and experience their place within the system of theoretical physics. Reason gives the structure to the system; the data of experience and their mutual relations are to correspond exactly to consequences in the theory. On the possibility alone of such a correspondence rests the value and the justification of the whole system, and especially of its fundamental concepts and basic laws. But for this, these latter would simply be free inventions of the human mind which admit of no a priori justification either through the nature of the human mind or in any other way at all. [...] the fictitious character of the principles [of mechanics] is made quite obvious by the fact that it is possible to exhibit two essentially different bases, each of which in its consequences leads to a large measure of agreement with experience. This indicates that any attempt logically to derive the basic concepts and laws of mechanics from the ultimate data of experience is doomed to failure. (1934, 165)

Hypothetico-deductivism thus claims that science starts with a hypothesis, deductively derives empirical consequences from the hypothesis, and tests these consequences using experiments and observations. Obviously, hypothetico-deductivism is in many ways opposed to inductivism: (i) While inductivists are generally wary of hypotheses, for hypothetico-deductivists hypotheses are

---

[12] See footnote 8.

the very starting point of the scientific enterprise. Any general law or theory will always remain hypothetical to a certain extent and can never definitely be proven true ("fictitious character of the principles"). (ii) Accordingly, hypothetico-deductivists deny that there is a universal inductive method with which laws and theories can be uniquely derived from experience ("any attempt logically to derive [...] is doomed to failure"). (iii) Instead, they emphasize the crucial importance of deductive inferences ("the data of experience and their mutual relations are to correspond exactly to [deductive] consequences in the theory. On the possibility alone [...] rests the value and justification of the whole system"). (iv) In a rationalist vein, hypothetico-deductivism emphasizes the contributions of the mind, of intuition and creativity, for the formulation of hypotheses (e.g., when Einstein speaks of the fictitious character of the principles of mechanics). (v) These intellectual contributions also underline the indispensable role of background knowledge and modeling assumptions for a hypothetico-deductive scientific methodology.

Some hypothetico-deductivists, in particular those in the Popperian tradition of falsificationism, are reluctant to ascribe probabilities to general laws or theories. But other hypothetico-deductive traditions are less dogmatic and allow for some kind of deductive confirmation of hypotheses, which may entail probabilities for these hypotheses. Still, such hypothetico-deductive confirmation fundamentally differs from the inductive verification envisioned by inductivists, at least by relying on deductive rather than inductive inferences.

As should be clear from the respective lists of characteristics, inductivism and hypothetico-deductivism are in many aspects irreconcilable (see Fig. 3). An inductive inference of general laws, for example, is incompatible with the requirement that these laws subsequently need to be deductively tested in an

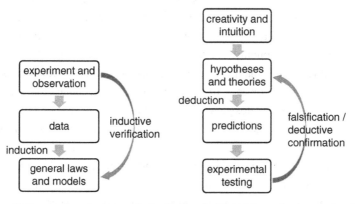

Figure 3 Comparison between inductivism (left) and hypothetico-deductivism (right).

attempt to falsify or corroborate them or with the view that in principle all laws always remain hypothetical. Thus, merely pointing out that science always involves inductive and deductive inferences falls short of showing how inductivist and hypothetico-deductive elements in scientific method can be synthesized.

As the most influential scientist of his age, Newton with his inductivist approach laid out in his "rules of reasoning" from the *Principia Mathematica* paved the way for inductivism. Similarly, Einstein's views helped establish hypothetico-deductivism as the dominant methodological framework of his time and subsequent decades. From a broader historical perspective, the quick dismissals of inductivism by modern commentators on big data approaches are astonishing. After all, inductivist approaches have to some extent been endorsed by some of the most successful scientists of all times including Newton, Lavoisier, and Ampère. This should be reason enough for inductivism not to be ridiculed or polemically labeled 'naïve.' Of course, inductivism has come under attack, for example, Newton and Ampère's version by Pierre Duhem in his main work *Aim and Structure of Physical Theory*, and the respective arguments merit careful examination. But the history of science gives the inductivist big data advocate ample opportunity to point to some of the greatest scientists of all times who at least paid lip service to inductivism.

3.3 The Inductivist Nature of Big Data Approaches

All the success stories of big data approaches described in the introduction exhibit the characteristics of inductivism of Section 3.2. For a brief discussion, let us illustrate this by means of the study of dermatologist-level classification of skin cancer, which was published in *Nature* a few years ago (Esteva et al. 2017).

Most importantly, everything starts with enormous amounts of data. The deep neural network for skin cancer classification was trained with approximately 130,000 skin lesion images showing over 2,000 diseases, which were arranged in a tree-structured taxonomy. The first level of the taxonomy classified the skin lesions as either benign, malignant, or nonneoplastic; the second level contained major disease classes for each of these categories (e.g., melanoma, which is malignant, or genodermatosis, which is nonneoplastic). Further levels contained subgroups of these disease classes. Since the correct classification of the images was known, it was a supervised learning problem (cf. Section 4.2). The data in this case were the images and in particular the individual pixels making up these images. These data all represent singular facts, which fits well with the finding from Section 3.2 that inductivism starts with statements of fact derived from observations and experiments.

Besides the data, the second crucial ingredient of big data approaches consists in various machine learning algorithms that are able to derive predictions from the data. In the present example, a deep neural network was pretrained at general object recognition and then fine-tuned on the data set of skin lesion images.[13] As I will argue in Section 4, deep neural networks, although having some superficial similarities with a hypothetico-deductive approach, actually implement a type of variational induction. The general model inferred from the data (i.e., the trained network) then allows classifying skin lesions, of which it is yet unknown whether they are malignant or benign. The predominant role of inductive, as opposed to deductive, inferences was identified in Section 3.2 as another one of the key features of inductivism.

As already mentioned and as I will show in detail in Section 4, deep learning algorithms follow a variational rationale that is typical for inductivist approaches. The decisive evidence for such algorithms does not consist in regularities or constant conjunction but rather in a systematic variation of circumstances while continuously tracking the impact that these variations have on the phenomenon of interest. The need for variation in the evidence is the main reason why such a huge amount of images was necessary to achieve classification skills comparable to that of a professional dermatologist. It is crucial that the evidence covers a substantial amount of the whole domain of naturally occurring human skin lesions (cf. Section 2.3). Clearly, mere enumerative evidence of the same or very similar lesions over and over again would not do the trick.

While the workings of machine learning algorithms may sometimes be opaque due to the sheer number of steps that are executed, there is no mystical element in how they achieve their results. In particular, there is little room for intuition or creativity – at least once the data are prepared and the algorithms are chosen.[14] Certainly, the authors of the skin cancer study needed substantial experience and also considerable ingenuity for adequately preparing the image data and choosing the correct algorithm, for example, the type of neural network and the number of hidden levels. But once everything was set up, no human intuition or creativity was required for formulating hypotheses about which image features are most suitable for classifying lesions. Such modeling was all done automatically by the algorithm.

---

[13] This type of approach is called *transfer learning* and is commonplace, since the training of machine learning models is usually time-consuming and difficult.

[14] A related issue, whether machine learning algorithms could display creativity, has recently become controversial and is now widely discussed under the term 'computational creativity' at the intersection of computer science, cognitive psychology, philosophy, and the arts.

Similarly, little domain-specific background knowledge or theoretical assumptions were required beyond the correct classification of the images. For determining malignancy of lesions, the employed deep learning algorithms used only pixels and disease labels as input, but did not require any further knowledge about the composition and structure of human skin, about the nature of cell growth, about possible triggers of malignancy, and so on. In particular, no knowledge was required about the mechanism linking the identified image features with malignancy. The data set of correctly classified images was sufficient for modeling and predictions.

It is usually the case that the results of data science can be improved by adding more data, in line with the inductivist idea that additional experiments and observations lead to the continuous refinement of causal knowledge, often by determining exceptions and clarifying the range of applicability. By contrast, many hypothetico-deductive approaches lack this incremental nature of knowledge increase and consider the disruptive falsification of hypotheses to be a crucial step in the scientific process. In typical machine learning algorithms such as the deep neural networks of the skin lesion study, falsification of hypotheses does not occur. Additional data in general will not falsify a specific neural network, but will provide an opportunity to further train the network and make the respective inferences ever more exact and reliable until the classification skills of a professional dermatologist are surpassed. The big data models developed by deep learning algorithms are not tested and falsified by new data but are gradually adapted to those data.

In summary, the example of skin cancer classification shows all the characteristics of an inductivist approach. Conversely, it is not amenable to a hypothetico-deductive interpretation. In view of such case studies, a crucial and indispensable task for an epistemology of data science is the rehabilitation of inductivism, which has largely been rejected in recent empirical science. While big data approaches are in general inductivist, the inductivist approach of data science will, of course, not be adequate for solving *all* scientific problems. Depending on the type of problem and the evidence that is available, a hypothetico-deductive approach may often be the better methodological choice (cf. Woodward 2011, 171–2). Furthermore, a hybrid method supplementing a general model-based approach with big data elements for those parts of the model, where theory is lacking but plenty of data is available, may prove fruitful in some scientific fields. Such hybrid models have been suggested, for example, in climate science (Knüsel et al. 2019; see also Northcott, Sect. 3).

4 Machine Learning as Variational Induction

4.1 Enumerative, Eliminative, and Variational Induction

Having shown that big data approaches are essentially inductivist, many details of such inductivism remain in the dark. Maybe the most important question in that respect will be addressed in this section, namely what kind of induction big data approaches implement. Induction, here, can be understood in a broad sense as any systematic, nondeductive inference from particular instances to general laws or models. The most fundamental distinction with respect to inductive methodology is between enumerative, eliminative, and variational induction. From a historical perspective, these three types seem the most prevalent. However, in view of the complex debates on the nature of induction reaching back to ancient science and philosophy, not everyone might agree with this assessment. For example, eliminative and variational induction are often conflated in the literature, but, as I will explain below, they are conceptually distinct and should be kept apart. Also, at least in principle, there remains the possibility that novel types of induction may emerge in the future.

According to the central thesis of this section, machine learning implements variational induction. Different lines of argument for this thesis extend through the whole Element and mutually support each other. Neither of them is fully conclusive by itself, but taken together a plausible picture emerges in support of the thesis. First, enumerative and eliminative induction are shown to be inadequate as a logical basis for machine learning (Sections 4.3.1, 4.3.2, 4.3.3). Second, crucial similarities between machine learning and variational induction are pointed out, in particular that both rely on difference making (Sections 4.2, 4.3.3). The third line of argument consists in a rational justification of variational induction by addressing all criticisms that have been raised against induction in general and variational induction in particular. In part due to lack of space, this line of argument has to remain sketchy (see, e.g., the framework proposed in Section 4.1.4 and the discussion of the problem of induction in Section 6.2). Finally, a fourth line of argument works out the consequences of identifying variational induction with the central type of induction in data science and shows that these consequences fit well with scientific practice (Sections 5 to 7).

4.1.1 Enumerative Induction

Enumerative induction is arguably the oldest and most familiar type of induction. As the term indicates, such inferences are established by an enumeration of instances that link similar events of a type A (e.g., the occurrence of a certain type of skin lesions) with similar events of another type C (e.g., the lethality of that type of skin lesions). Given a sufficient number of observations where A and C co-occur and lacking instances in which C is absent in the presence of A, it is justified to conclude that there is a general dependency between A and C. For example, observing that a certain type of skin lesions is lethal in a number of patients leads to the conclusion that these skin lesions kill in general.

It is unclear, however, exactly how many instances are necessary to justify an inference to a general law. In fact, this constitutes one of the major objections against enumerative induction, which was already raised by David Hume (1748) in his notorious critique of induction and subsequently reiterated by many others.[15] Sometimes, successful enumerative inferences are based on just a few instances, while often even a large number of instances may lead to erroneous conclusions. One could also frame this issue as the problem of the measure of confirmation for enumerative induction. Indeed, the mere number of instances on which an inductive inference is based does not constitute an adequate measure of confirmation. After all, the number of instances that is deemed satisfactory for a reliable inductive inference often depends on the specific context. At least to my knowledge, a systematic, quantitative approach, how an adequate measure of confirmation could be derived from the respective context, does not exist. Another crucial and related objection against enumerative induction regards the absence of a criterion to distinguish meaningful from merely accidental conjunctions of events, i.e., between evidence that warrants reliable predictions and evidence that does not.

Nevertheless, in epistemology and also in the sciences, enumerative induction is held by many distinguished scholars to be an important, if not the crucial, type of induction:

> The most ancient form of induction, the archetype of this family, is enumerative induction, or induction by simple enumeration. [...] Examples are readily found in Aristotle. Traditionally, enumerative induction has been synonymous with induction, and it was a staple of older logic texts to proceed from deductive syllogistic logic to inductive logic based on the notion of enumerative induction. [...] The actual inductive practice of science has always used enumerative induction, and this is not likely to change. For example, we believe all electrons have a charge of -1.6×10^{-19} Coulombs, simply because all electrons measured so far carry this charge. (Norton 2005, 10–12)

Certainly, enumerative induction seems to play a role in everyday reasoning. One usually expects that a relationship that has been observed many times will

---

[15] See Vickers (2018) and references therein.

continue to hold in the future. Enumerative reasoning in a more general sense, i.e., various kinds of reasoning based on the enumeration of similar instances, also lies at the heart of many conceptual analyses in epistemology (e.g., the frequency interpretation of probability or regularity conceptions of scientific laws and of causation).

However, I will argue further below that, instead of being the core of inductive methodology, enumerative induction does not even constitute an important type of induction. In fact, enumerative induction should actually be understood in terms of variational induction when controlling for possibly relevant circumstances, the influence of which has so far been neglected. Thus, alleged enumerative induction often constitutes an application of the method of agreement of variational induction.

4.1.2 Eliminative Induction

The second type of induction to be discussed here is *eliminative induction*. This type relies on the assumption that an exhaustive set of hypotheses or theories about a phenomenon is available. One after another, all false hypotheses or theories can be eliminated until a single one remains that best represents the phenomena. This elimination in general proceeds deductively via *modus tollens* by showing that certain deductive consequences of a hypothesis or theory are in contradiction to experience.

A classic example for eliminative induction concerns the discovery of the causes of childbed or puerperal fever by Ignaz Semmelweis in the nineteenth century (Bird 2010).[16] In two maternity wards of the General Hospital in Vienna in the 1840s, mortality rates widely differed. Whereas in one ward only about 2 to 3 percent of the women died when giving birth, in the other, mortality rates were much higher, in some years even above 10 percent. Ignaz Semmelweis was a young doctor at the time, employed at one of the wards and determined to solve the puzzle of this remarkable situation. Based in part on his own detailed documentation, Semmelweis formulated a number of possible hypotheses regarding the cause of the high mortality rates. For example, he observed that women in the first ward delivered on their backs, while those in the second ward delivered on their sides. Thus, he arranged for the women in the first ward to also deliver on their sides, but with no effect on mortality – thereby eliminating the corresponding hypothesis. After many such eliminations, Semmelweis hypothesized that the women in the second ward contracted an infection from medical students, who were trained in that ward and who as part of their

---

[16] This example has also been used to illustrate rivaling accounts of scientific inference (e.g., Hempel 1966, Ch. 2; Scholl 2013).

education also assisted in autopsies. By comparison, in the first ward, midwives were trained who did not take part in autopsies. Semmelweis could show that consequences of this last hypothesis largely turned out to be true. In particular, having the medical students wash their hands after autopsies with chlorinated water, a disinfectant, significantly lowered the mortality rate in the second ward.

Eliminative induction has been criticized for various reasons. One issue regards terminology as it may not constitute an inductive method at all. Indeed, it resembles much more closely a hypothetico-deductive approach, where hypotheses are postulated and their deductive consequences are examined. Obviously, the main additional premise in comparison with conventional hypothetico-deductivism is that an exhaustive set of hypotheses is available, among which the true or at least a highly probable hypothesis can be found. In general, this premise seems impossible to establish, and it is already wildly implausible in the case study of childbed fever. In fact, the eventual 'true' hypothesis was not even among the original set that Semmelweis considered, but was suggested to him by the death of a close colleague who succumbed to a disease with very similar symptoms as childbed fever after an infection contracted during a postmortem examination (Bird 2010, 6). Not least because the premise of an exhaustive set of hypotheses is for most applications unrealistic, this premise does little work in eliminative induction, and thus the classification of eliminative induction as a hypothetico-deductive approach seems reasonable.

4.1.3 Variational Induction

Let us finally turn to *variational induction*.[17] Representative of this type are John Stuart Mill's five canons of induction (1886).[18] Arguably, the quintessential method of variational induction is the method of difference. As an example illustrating this method, one can learn that a light switch is the cause or at least causally related to a corresponding light, by comparing a situation where the switch is turned off and the light is off with another situation in which the switch is turned on and the light is on. Intuitively, one also has to ensure that nothing else that might be relevant has changed in the meantime; for example, that no one has activated other switches. Mill formulates the method of difference as follows:

---

[17] This terminology, which is not standard, is borrowed from Russo (2007, 2009), who has emphasized a 'rationale of variation' underlying much of our inductive reasoning. In previous papers, I used the term 'eliminative induction' for 'variational induction' though I was always careful to point out that there are two very distinct types of eliminative induction (e.g., Pietsch 2014).

[18] Bacon's method of tables is an important predecessor of Mill's methods.

If an instance in which the phenomenon under investigation occurs, and an instance in which it does not occur, have every circumstance save one in common, that one occurring only in the former, the circumstance in which alone the two instances differ, is the effect, or cause, or a necessary part of the cause, of the phenomenon. (1886, 256)

Another important method in Mill's approach is the method of agreement:

If two or more instances of the phenomenon under investigation have only one circumstance in common, the circumstance, in which alone all the instances agree, is the cause (or effect) of the given phenomenon. (1886, 255)

Both methods infer a causal relationship between a phenomenon and its circumstances by relying on variational evidence, i.e., on evidence that tracks changes in a phenomenon resulting from systematic variations of circumstances. Essentially, they look for *difference makers* among the circumstances, i.e., for those circumstances that have the largest or at least a substantial impact on the phenomenon of interest. Note that Mill's requirement for the method of difference that the instances differ in only one circumstance is a version of the so-called *homogeneity condition*, which is typical for variational induction. It ensures that no other relevant circumstances may change except the one that is explicitly considered. In the above example of the light switch, homogeneity is realized by the condition that only the explicitly considered light switch but, for example, no other switches have changed between the two considered situations.

Variational induction, in particular the version of Mill's methods, is beset with a number of problems, which may have contributed to its relative neglect in the epistemological literature. As an example, it is implausible that the conditions for Mill's formulation of the method of difference can ever be met, namely that two instances differ in only a single circumstance. Instead, there will always be myriads of circumstances in the universe that keep changing from one instance to the other. Also, Mill's methods are not suitable to determine causal factors, when several rather than just a single circumstance determine a phenomenon, and alternative causes. More exactly, a causal factor is a circumstance that requires other circumstances to be present in order to have an impact on the phenomenon (e.g., a burning match that requires combustible material to start a fire). An alternative cause is a circumstance that requires other circumstances to be absent in order to have an impact on the phenomenon (e.g., a lawn sprinkler system that should be used to keep the lawn wet only if it does not rain). Recent research has shown that there are somewhat plausible solutions for how causal factors and alternative causes can be taken into account in a framework of variational induction (Mackie 1967; 1980, Appendix; Graßhoff & May 2001; Baumgartner

& Graßhoff 2003; Schurz 2014, 182–7; Pietsch 2014; 2016b; 2019, Sect. 4.3; Baumgartner & Falk 2019).

There are a number of fundamental differences between variational and enumerative induction. While in enumerative induction the focus lies on regularities of events, in variational induction it is on changes in circumstances. Consequently, the decisive type of evidence and the measure of confirmation differ for both inductive methods. In enumerative induction, confirmation is mostly thought to increase with the number of observed co-occurrences of two phenomena A and C. By contrast, confirmation in variational induction increases with the variety of evidence, i.e., with observing as many different situations in terms of changing circumstances as possible. Relatedly, fully identical instances, i.e., those with exactly the same circumstances, are confirmatory of a generalization according to enumerative induction, but they are not confirmatory according to variational induction. While negative instances ¬C play only a destructive role in enumerative induction falsifying the respective generalization, they are constructively taken into account in variational induction leading to the refinement of causal knowledge. All these differences imply, not surprisingly, that whether and how inductive inferences can be justified is substantially different between enumerative and variational induction.

Many proponents of variational induction have been quite vocal about their rejection of enumerative induction, including influential figures like Francis Bacon, John Stuart Mill, or John Maynard Keynes. The first writes that "the induction [...] which proceeds by simple enumeration is a childish affair, unsafe in its conclusions, in danger from a contradictory instance, taking account only of what is familiar, and leading to no result" (1620, 21). Mill agrees: "[Enumerative induction] is the kind of induction which is natural to the mind when unaccustomed to scientific methods. [...] It was, above all, by pointing out the insufficiency of this rude and loose conception of Induction that Bacon merited the title so generally awarded to him of Founder of the Inductive Philosophy" (1886, 204). The economist John Maynard Keynes in his *Treatise on Probability* emphasizes the importance of variation as opposed to regularity: "by emphasizing the number of instances Hume obscured the real object of the method. [...] The variety of the circumstances [...] rather than the number of them, is what seems to impress our reasonable faculties" (1921, 233–4).

Let me also briefly address how variational induction differs from eliminative induction. This is particularly important because many consider Mill's canons to be a main example of eliminative induction. For example, John Norton writes: "Mill's canons provide some of the best known examples of eliminative induction. He labelled them 'methods of elimination' since they are intended to enable one to eliminate all but the true causes out of the range of possible causes for a given phenomenon" (1995, 30). And indeed, Mill himself claims that his methods of agreement and of difference constitute an eliminative approach: "[The term 'elimination'] is well suited to express the operation [. . .] which has been understood since the time of Bacon to be the foundation of experimental inquiry: namely, the successive exclusion of the various circumstances which are found to accompany a phenomenon in a given instance, in order to ascertain what are those among them which can be absent consistently with the existence of the phenomenon" (1886, 256).

But, as I want to argue, the parallels are rather superficial. Importantly, in variational induction, *circumstances* are eliminated, for example, by showing their irrelevance to a phenomenon, while in eliminative induction competing *hypotheses* are eliminated. The circumstances whose influence is examined in variational induction do not per se constitute competing hypotheses. Rather, these circumstances are typically combined to form a Boolean expression, i.e., a combination of several circumstances connected by the logical operators 'and' and 'or' to account for causal factors and alternative causes. Such Boolean expressions in their entirety may then be interpreted as potential causes of a given phenomenon. Because individual circumstances may interact in various ways with other circumstances, they should not be interpreted as competing hypotheses, which are successively eliminated.

Relatedly, circumstances are judged by variational induction in terms of completely different categories than competing hypotheses in eliminative induction. Circumstances can be classified as necessary or sufficient or, alternatively, as causally relevant or causally irrelevant (cf. the alternative framework sketched below). By contrast, competing hypotheses are either falsified or provisionally upheld by eliminative induction. It would be a category mistake to assume that circumstances are falsified in variational induction or that hypotheses are identified as causally relevant or irrelevant in eliminative induction. Furthermore, the negation of a circumstance can turn out to be causally relevant in variational induction, whereas false hypotheses do not play a constructive role in eliminative induction. Also, circumstances can undergo gradual changes rather than being only absent or present, while nothing analogous can be said about competing hypotheses.

Finally, the spectrum of different methods in variational induction has no analogue in eliminative induction. For example, the method of difference does not aim at elimination or falsification, but directly establishes a causal relationship between groups of variables. Something similar could be said about Mill's method of concomitant variation, which tracks causal relationships between gradual changes in different variables. By contrast, eliminative induction can establish the truth or adequacy of a hypothesis only by elimination of all potential rivals, but never directly in a single inference as in the case of the methods of difference and concomitant variation.

One might object that also in variational induction a hypothesis is examined, namely the hypothesis that a certain Boolean expression of circumstances constitutes a cause of a phenomenon. Indeed, in this manner, it is possible to map variational induction onto eliminative induction in principle. However, the hypothetico-deductivism of eliminative induction provides an unnatural perspective on variational induction, not least because the former is about deductive hypothesis evaluation, while the latter is concerned with the inference of general laws from experience. Also, the methods of variational induction (e.g., the method of difference) do not examine any hypotheses directly, but rather examine the role of individual circumstances of a phenomenon.

Let me briefly illustrate these fundamentally different perspectives by the example of the light switch. A researcher using variational induction would first try to determine all potentially relevant circumstances for the considered light switch in a given context, such as different switches or the presence of various electric cables. She would then vary individual circumstances (e.g., a specific switch) while at the same time trying to ensure homogeneity by holding all other potentially relevant circumstances constant (i.e., the states of all other switches and the presence or absence of the respective electric cables). If homogeneity holds, relevance or irrelevance of the specific switch to the considered light can be inferred with respect to a context determined by the states of the other potentially relevant circumstances. In this manner, relevance and irrelevance information about the various circumstances can be gathered constantly, expanding the overall picture of how the different switches and electric cables are causally related to the considered light.

For comparison, a researcher employing eliminative induction might also start by determining all potentially relevant circumstances. But she would then formulate hypotheses that take into account *all* of the potentially relevant circumstances, for example, that all of the switches except one are off and that all electric cables are connected. She would need to somehow ensure that the set of hypotheses is exhaustive and then proceed to falsify hypotheses until only a single one remains. Obviously, homogeneity plays no role in this procedure, the researcher learns nothing about the relevance or irrelevance of individual circumstances, and the eliminative approach lacks the incremental information gathering of variational induction.

Thus, there are plenty of reasons to hold eliminative and variational induction apart, even though the two have mostly been conflated in the past.

4.1.4 An Alternative Variational Framework[19]

As another example of a framework of variational induction that may solve some of the problems raised by Mill's approach, I have elsewhere suggested a number of fundamental definitions, which shall be very briefly presented in the following (Pietsch 2016b; see also Pietsch 2014; 2019, Sect. 4.3).

The framework is based on two fundamental methods, the *method of difference* to determine causal relevance of a circumstance for a phenomenon and the *strict method of agreement* to determine causal irrelevance:

> Method of difference: If two instances with the same background B are observed, one instance, in which circumstance A is present and phenomenon C is present, and another instance, in which circumstance A is absent and phenomenon C is absent, then A is causally relevant to C with respect to background B, iff B guarantees homogeneity.

> Strict method of agreement: If two instances with the same background B are observed, one instance, in which circumstance A is present and phenomenon C is present, and another instance, in which circumstance A is absent and phenomenon C is still present, then A is causally irrelevant to C with respect to background B, iff B guarantees homogeneity.

Here, an instance has a background or context B, if a given set of circumstances, by which B is defined, are all present – or more generally if a given set of circumstantial variables has certain predetermined values. The background usually refers to a forbiddingly large number of circumstances or variables, which due to their sheer number cannot be made explicit. It may include events at distant places and in the early history of the universe.

Homogeneity essentially captures the intuition that all circumstances that are causally relevant to the examined phenomenon must be held fixed for the comparison between instances in the above methods, except those circumstances whose influence is explicitly examined (e.g., Holland 1986). More precisely:

> Context B guarantees 'homogeneity' with respect to the relationship between A and C, iff the context B is thus defined that only circumstances that are causally irrelevant to C can change, (i) except for A and (ii) except for

---

[19] This brief subsection is somewhat technical and presupposes deeper familiarity with epistemological discussions about variational induction. A reader interested only in the main argument of the Element can easily skip the subsection. It is included for those familiar with typical objections against variational induction to serve as an indication that at least some of these issues may be solvable.

circumstances that are causally relevant to C in virtue of A being causally relevant to C.

The second exception allows for circumstances to change that colloquially speaking lie on a causal chain through A to C or that are effects of circumstances that lie on this causal chain. The above explication implements the aforementioned intuition behind the notion of homogeneity that factors in the background B that are causally relevant to the examined phenomenon C may not change or that only circumstances that are causally irrelevant are allowed to change.

What is lacking so far is the explicit link from the method of difference determining causal relevance and the strict method of agreement determining causal irrelevance to the notions of causal factors and alternative causes as defined in Section 4.1.3. This link is established by the following definitions:

> A is a 'causal factor' for phenomenon C with respect to background B, iff there exists an X such that A is causally relevant to C with respect to B & X and causally irrelevant to ¬C with respect to B & ¬X (i.e., C is always absent in B & ¬X).

> A is an 'alternative cause' to C with respect to background B, iff there exists an X such that A is causally relevant to C with respect to a background B & ¬X, but causally irrelevant to C with respect to a background B & X (i.e., C is always present in B & X).

The formulation B & X means that the context or background is determined by B plus the additional condition of state X being present, while B & ¬X means that the context is determined by B plus the additional condition of state ¬X being present (i.e., X being absent). Thus, both causal factors and alternative causes can be defined based merely on causal relevance and causal irrelevance. In other words, alternative causes and causal factors can be determined based merely on the method of difference and the strict method of agreement. Based on the notions of causal factors and alternative causes, the aforementioned Boolean expressions of circumstances can be constructed and thus a link to John L. Mackie's notion of an inus-condition can be established, by which the general concept of 'cause' is essentially explicated as a number of alternative causes, each comprising different causal factors (1980, Ch. 3).

While the above framework deals with causal relationships that are formulated in terms of the absence or presence of circumstances, the framework can be extended to functional relationships as well, i.e., dependencies between continuous variables. Indeed, Mill's canons include the method of concomitant variation, which also derives functional causal relationships (cf. also Pietsch 2014, Sect. 3.4).

4.2 Machine Learning Algorithms

Data sets that are so large that they cannot be handled by pencil and paper or by the human mind have to be analyzed by information and computation technologies. For this reason, the concepts of big data and of artificial intelligence as well as machine learning are inextricably linked.

4.2.1 Supervised and Unsupervised Machine Learning

The terms 'machine learning' and 'artificial intelligence' are often used in similar ways. And certainly, there are no true definitions that must necessarily be associated with these terms. Rather, it is partially a matter of conventional choice and pragmatic usefulness as to which definitions one prefers.

In the following, 'artificial intelligence' designates the whole field of artificially replicating various kinds of human intelligence as well as developing further 'unhuman' types of intelligence that may transcend the possibilities of human intelligence. Artificial intelligence encompasses in particular abstract reasoning as well as learning from experience. Of course, the methods and approaches of artificial intelligence need not always correspond to the methods and approaches with which the human mind proceeds (cf. Russell & Norvig 2009, Preface, Ch. 1.1; Clark 1996).

'Machine learning' forms part of artificial intelligence. Machine learning, as it is defined here, is restricted to algorithms of learning from data. This somewhat narrow interpretation is largely in accordance with the types of algorithms that are habitually considered to belong to machine learning, such as support vector machines, decision trees, neural networks, or various clustering algorithms (e.g., Hastie et al. 2001; Flach 2012; Ghani & Schierholz 2017). In many fields, machine learning approaches have replaced rule-based approaches to artificial intelligence. The latter answer questions based on a set of rules, which usually take the form of if/ then clauses and are not learned but are predefined. For example, in machine translation these rules would include basic grammatical rules for analyzing and constructing sentences. Such rule-based approaches require considerable efforts and costs for developing and maintaining the rules (Ghani & Schierholz 2017, 148).

All machine learning algorithms aim to predict one or more target variables based on a set of predictor variables. The predictions are learned from data sets comprising many instances that link different values of the considered variables. A fundamental distinction is between *supervised* and *unsupervised* machine learning. Two aspects in which these types of learning differ are usually highlighted. First, in supervised learning the target variable, which is to be predicted, is specified, while in unsupervised learning no target variable is predetermined and various patterns in the data may be picked up by the algorithms (Ghani & Schierholz 2017, 151–2). Second, the data structure differs in that supervised learning starts from a training set of input–output pairs, where the output corresponds to the value of a target variable and the input to corresponding values of the predictor variables. By contrast in unsupervised learning, the data by definition does not have such an input–output structure, since a distinction between predictor and target variables is not predetermined, but suggested by the algorithm (Russell & Norvig 2009, 694–5).

Widely used supervised learning algorithms are support vector machines, decision trees, random forests, neural networks, or k-nearest neighbors, but also more conventional statistical methods such as regression analysis. Typical unsupervised learning algorithms are, for example, various clustering algorithms such as k-means clustering or mean-shift clustering, principal component analysis, or association rules (e.g., Ghani & Schierholz 2017, Sect. 6.5; Ng & Soo 2017, Sect. 1.2).

The crucial question to be answered in the remainder of the present section concerns what kind of induction the abovementioned machine learning algorithms implement. I will show for several of the most widely used and most successful types of machine learning algorithms that they rely on a rationale of difference making, i.e., they all implement a version of variational induction.

4.2.2 Decision Trees

Decision trees are a type of supervised learning and are among the "simplest and yet most successful forms of machine learning" (2009, 697),[20] as Stuart Russell and Peter Norvig write in the leading modern textbook on artificial intelligence:

> In many areas of industry and commerce, decision trees are usually the first method tried when a classification method is to be extracted from a data set. One important property of decision trees is that it is possible for a human to understand the reason for the output of the learning algorithm. [. . .] This is a property not shared by some other representations, such as neural networks. (2009, 707)

Like most machine learning algorithms, decision trees can be considered as complex functions mapping several input or predictor variables on an outcome or target variable (which constitutes the 'decision'). Decision trees employ what in the machine learning literature is called a "greedy divide-and-conquer" strategy.

---

[20] For an early discussion from a philosophy of science perspective, see Gillies (1996, Sect. 2.3). In parts of the machine learning literature, these algorithms are somewhat more aptly termed 'classification and regression trees,' since they can in general be used for those two tasks. Of course, decision trees weighing between different options based on answers to predetermined questions also yield classifications (cp. the example in Fig. 4). Thus, strictly speaking, decision trees are a special case of classification trees aimed at decision making.

According to this strategy, it is first determined which input variable "*makes the most difference* to the classification of the example," i.e., to the outcome variable (Russell and Norvig 2009, 700, my emphasis). For each value or set of values of that input variable a subproblem results. For each subproblem again the input variable among the remaining input variables is sought that makes the most difference, and so on (2009, 700). Thus, the decision process proceeds on a tree-like graph, where different branches are selected depending on the values of the respective input variables, which constitute the nodes of the graph. Each of these branches terminates with a leaf, which ultimately fixes the value of the outcome variable. A widely used example of a decision tree concerns the deliberation process of whether one should wait at a restaurant or not depending on variables such as the number of patrons currently in the restaurant, whether one is hungry, or whether it is a weekend (see Fig. 4).

The variable that makes the most difference is the variable that conveys the most information with respect to classification. Information can be measured in terms of the Shannon entropy, which according to information theory describes the uncertainty of a random variable. More exactly, the algorithm proceeds as follows for a classification problem regarding an outcome variable C, which can take on the values $c_1, \ldots, c_n$. Given that these values appear with relative frequencies $p(c_1), \ldots, p(c_n)$ in the training set, the Shannon entropy is then calculated as $H(C) = -\sum_i p(c_i)log_2 p(c_i)$. The Shannon entropy is maximal, when all outcomes are equally likely, and minimal in the case of perfect classification, i.e., if the probability of a single c_x equals one, and the probabilities for all other c's are zero. In the aforementioned

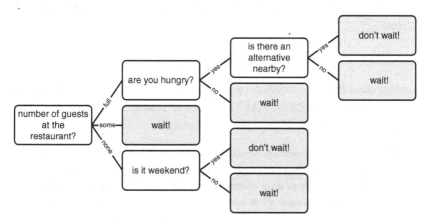

Figure 4 Example of a decision tree determining whether one should wait to be seated at a restaurant. White boxes designate input variables and grey boxes designate values of the outcome variable (cf. Russell & Norvig 2009, 702).

example, the outcome variable C is binary, whether one decides to wait at the restaurant or not.

One can then introduce a classification based on a variable A_X with possible values $x_1, \ldots, x_k$. In the example, A_X could be the number of patrons at the restaurant or whether it is a weekend. Again, $p(x_j)$ and the conditional probabilities $p(c_i|x_j)$ can be determined on the basis of relative frequencies in the training set. Accordingly, a conditional Shannon entropy can be calculated: $H(C|A_X) = \sum_j p(x_j) H(C|A_X = x_j) = -\sum_j p(x_j) \sum_i p(c_i|x_j) log_2 p(c_i|x_j)$. The so-called information gain is defined as $H(C) - H(C|A_X)$. It is always positive or zero and quantifies the improvement of the classification $C|A_X$ with respect to just C. The information gain is maximal, namely $H(C)$, for a perfect classification, i.e., a classification where all $p(c_i|x_j)$ are either 1 or 0. In this way, the A_X can be determined that exhibits the largest information gain of all A, resulting in a number of subtrees corresponding to the different values of the parameter A_X.

The whole procedure is then repeated for every subtree. Typical stopping criteria include that the instances included at each leaf must all have the same values for the outcome variables, that each leaf must contain a predetermined number of instances or that further branching does not improve the classification beyond a certain threshold (Ng & Soo 2017, 81).

In the case of a perfect classification, the tree structure resulting from the algorithm yields an expression of necessary and sufficient conditions for each value of C: for example, iff $(A_1 = y_1 \wedge A_2 = y_2) \vee (A_1 = y_3 \wedge A_3 = y_4)$, then $C = c_1$. These correspond exactly to the type of Boolean expressions discussed in Section 4.1.3 in connection with variational induction. A decision tree usually yields fairly reliable predictions, whenever it manages to at least approximate some of the actual necessary and sufficient conditions for the phenomenon of interest.

4.2.3 Neural Networks

Another class of highly successful algorithms is so-called deep neural networks or deep learning algorithms.[21] As Yann LeCun, Yoshua Bengio, and Geoffrey Hinton write in a recent review article in *Nature*:

> Deep learning [...] methods have dramatically improved the state-of-the-art in speech recognition, visual object recognition, object detection and many other domains such as drug discovery and genomics. [...] Deep convolutional nets have brought about breakthroughs in processing images, video,

---

[21] Much of the following is taken from the first textbook on *Deep Learning*, written by some of the major figures in the field: Ian Goodfellow, Yoshua Bengio, and Aaron Courville (2016). Wikipedia also was a helpful source.

speech and audio, whereas recurrent nets have shone light on sequential data such as text and speech. (LeCun et al. 2015, 436)

Both deep convolutional neural networks and recurrent neural networks are specific, widely used types of deep learning. The former are particularly suitable for tasks processing a grid of values such as in image recognition. One useful feature is their scalability, e.g., from smaller to larger images. By contrast, recurrent neural networks are especially good at dealing with sequential data, e.g., temporal sequences in language processing (Goodfellow et al. 2016, 363).

The architecture of artificial neural networks is vaguely inspired by the human brain. In analogy to the human brain, the fundamental building blocks of neural networks are artificial neurons, which are connected with each other. As is well studied, biological neurons communicate electrochemically. They process an input from other neurons to generate an electric output signal, which they then feed to still further neurons via so-called axons. In the human brain, neurons generally become activated when the input signal surpasses a certain threshold, and after activation, they send a spike to other neurons, potentially activating those other neurons in turn. In the brain, the neurons are often organized in several subsequent layers (e.g., in the visual cortex, which is one of the best understood parts of the brain).

Efforts to computationally model the processes in the brain have developed in parallel with an increased understanding of the brain. The first substantial attempt at such modeling was the so-called 'perceptron,' which was invented by the American psychologist Frank Rosenblatt at Cornell Aeronautical Laboratory in the late 1950s.[22] Rosenblatt developed a number of algorithms for neural networks, which were eventually implemented on some of the first computers. Also, an image recognition machine, the 'Mark I Perceptron,' was built consisting of an array of 20×20 photocells that were connected with a number of response cells. The adaptive weights of the connections between photocells and response cells were realized by potentiometers that could be controlled using electric motors.

Mathematically, the perceptron can be modeled in terms of a function that maps an input vector x to an output value $f(x)$. In the simplest case the output is a binary classifier (see Fig. 5):

---

[22] Rosenblatt writes in the preface to his *Principles of Neurodynamics*: "For this writer, the perceptron program is not primarily concerned with the invention of devices for 'artificial intelligence', but rather with investigating the physical structures and neurodynamic principles which underlie 'natural intelligence'. A perceptron is first and foremost a brain model, not an invention for pattern recognition. As a brain model, its utility is in enabling us to determine the physical conditions for the emergence of various psychological properties" (1962, vi).

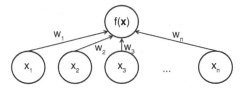

Figure 5 Model of a simple perceptron with inputs x_i, corresponding weights w_i, and output $f(x)$.

$$f(x) = \begin{cases} 1 \text{ if } wx + b > 0 \\ 0 \text{ otherwise} \end{cases}$$

Here, the vector w denotes the respective weights, with which the various input values are taken into account for the output, as realized by the potentiometers in the Mark I perceptron, and b is the bias shifting the threshold, above which the classifier yields a positive result.

Crucial restrictions of simple perceptron models turned out to be the linear nature of the connections between input and output as well as the lack of depth of the network consisting of only an input and an output level. Indeed, a perceptron of the above design is not able to learn important types of functions, e.g., XOR,[23] as pointed out in the notorious book *Perceptrons* by two further AI pioneers, Marvin Minsky and Seymour Papert. This book, with its rather pessimistic outlook, arguably played a crucial role in the subsequent decline in interest in neural networks over a period of approximately a decade. Remarkably, periods of renewed and declining interest occurred several times in the history of AI. Computer scientists speak rather poetically of the winters of AI. Goodfellow et al. make out three different periods of development of neural networks, the first in the 1940s through the 1960s in the field of cybernetics, then in the 1980s and 1990s under the term 'connectionsm,' and in the last decade or so, the term 'deep learning' has been used (2016, 12).

In recent times, so-called deep neural networks have proven to be very effective for a large range of applications. In contrast to the simple perceptron model as employed in the Mark I, such networks are arranged in a number of layers, which subsequently process and transform the input data. Because the variables in the intermediary layers cannot be directly observed, these layers are also referred to as hidden layers (see Fig. 6). But hidden layers alone do not solve the XOR problem. Thus, a nonlinear mapping between subsequent

---

[23] The XOR-function is defined as follows: $f(1,1) = f(0,0) = 0; f(1,0) = f(0,1) = 1$.

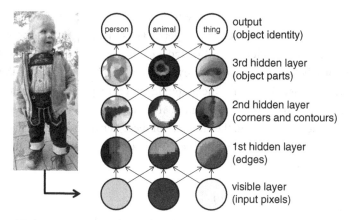

Figure 6 Schema of a neural network with circles representing nodes arranged in subsequent layers and the arrows representing the weighted connections between the nodes (cf. Zeiler & Fergus 2014; Goodfellow et al. 2016, 6).

layers was a further crucial element in dealing with the shortcomings of some early realizations of the perceptron.

Essentially, these deep neural networks are just computational representations of mathematical functions: "A multilayer perceptron is just a mathematical function mapping some set of input values to output values. The function is formed by composing many simpler functions. We can think of each application of a different mathematical function as providing a new representation of the input" (Goodfellow et al. 2016, 5). A crucial advantage of deep learning algorithms is that they can represent almost any functional relationship with sufficient accuracy in contrast to traditional statistical modeling tools such as linear regression, which are typically limited to specific families of functions (e.g., linear relationships).

In current research, a considerable number of algorithms for artificial neural networks are developed, many of them specialized for certain types of input data (e.g., the before-mentioned convolutional neural networks and recurrent neural networks). In the following, I will concentrate on so-called feedforward deep neural networks, which are widely considered "the quintessential example of a deep learning model" (2016, 5).

Almost any machine learning algorithm relies on a *data set* that is represented by a computational *model*. A *cost function* measures the discrepancy between the actual data and the corresponding results of the model. Finally, an optimization procedure improves the model, such that the cost function is minimized to an acceptable extent (2016, 149–50). Deep feedforward networks are multilayer artificial neural networks with an input layer representing those variables that correspond to the input data of the data set. An arbitrary number of hidden layers follow, where direct links only exist between variables in subsequent layers. Thus, from a graph-theoretical perspective such models are directed acyclic graphs, lacking feedback connections between the various levels. Finally, the variables of the last layer, which is usually termed the output layer, correspond to the output data of the data set (2016, 163–6; LeCun et al. 2015; see also Fig. 6).

Mathematically speaking, the structure of these networks consists in a chain of nested mathematical functions (i.e., a hierarchical sequence of functions which take each other as input variables):

$$f^{(n)}\left(\ldots \; f^{(3)}\left(f^{(2)}\left(f^{(1)}(x) \right) \right) \ldots \right)$$

Here, the upper index denotes the respective layer. For example, $f^{(1)}$ determines the values on the first hidden layer, $f^{(2)}$ the values on the second layer, and so on with n being the overall depth of the network. The chain of functions has the Markov property, i.e., the values in every layer depend only on the values in the preceding layer. Note again that at least some of the functions f must be nonlinear so that the network is able to represent nonlinear functions such as XOR.

A typical choice for a function f connecting subsequent layers consists in a linear transformation and a subsequent application of a nonlinear activation function, which is usually applied elementwise. Various nonlinear activation functions have been tried in the past. Currently, the default recommendation is to use the so-called rectified linear unit, often abbreviated as ReLU, which yields the maximum of either the input value or zero: $g(z) = max\{0, z\}$ (Goodfellow et al. 2016, 168).[24] Because ReLU and similar functions are not differentiable at all points, they were originally rejected out of what in hindsight seems to be mostly superstition. But they turned out to yield good results in many practical applications. Thus, we have as a typical mapping from one layer to the next:

$$f(x) = max(0, Wx + b)$$

The maximum is taken elementwise, i.e., the result is again a vector, denoting the values of the input variables on the subsequent level. W is a matrix linearly transforming the vector x, and b is the bias adjusting the offset of the function. The entries of the matrix W denote the strengths of the connections between the different nodes or variables in subsequent layers. For example, the value at

---

[24] Often, the choice of activation function is largely irrelevant: "New hidden unit types that perform roughly comparably to known types are so common as to be uninteresting" (2016, 190).

position (1,1) in the matrix determines the strength of the connection between the first nodes in subsequent layers. Obviously, the number of rows and lines of the matrix determine the number of nodes in subsequent levels.[25]

The weights *W* determining the strength of the connection between nodes in subsequent layers as well as the bias *b* of the various layers are the model parameters that are optimized in the machine learning process. In typical neural networks, the number of parameters is usually quite large, not least because the number of input variables is often already substantial. For example, in image processing, the color values of every pixel can be taken as input variables. In the beginning, the parameters are initialized, often to some small random values (2016, 172). Then, an optimization procedure is applied, trying to change the parameters in a way such that the output values of the model are as close as possible to the corresponding values in the data set.[26]

The standard optimization procedure in machine learning is gradient descent. For this procedure, a suitable cost, loss, or error function is chosen measuring the difference between the model output and the corresponding 'desired' values from the data set. Then, the gradient of this cost function is determined with respect to the parameters of the model. The gradient essentially determines how the parameters of the model must be changed in order to minimize the cost function, i.e., in order for the model to achieve better results. From a computational perspective, this optimization procedure is implemented in small steps. The gradient is calculated locally at the current values of the parameters, and then the values of the parameters are adjusted to a small extent, expressed in terms of a *learning rate α*, such that the impact on the cost function is optimized. The procedure is repeated for the adjusted values of the parameters again and again until the cost function reaches a (local) minimum, where the gradient is zero and further changes in the parameters do not lead to any improvement in the cost function. Such gradient descent can be visualized as a walk in a multidimensional landscape, where an imagined hiker tries to get down from a mountain as fast as possible (cf. Fig. 7).

Gradient descent determines which input variables are difference makers and which are irrelevant for classification. For the latter, the connection to subsequent layers is erased by setting the corresponding weights to zero.[27] Deep neural networks, like many other machine learning algorithms, rely on

---

[25] The exact network architecture is mostly determined by trial and error: "The ideal network architecture for a task must be found via experimentation guided by monitoring the validation set error" (2016, 192).

[26] At least in supervised learning. In unsupervised learning, one would expect that the output captures relevant aspects of the examined phenomenon.

[27] In practice, many weights will be set close to zero and the neural network more often than not will only approximate the 'true' difference-making relations. This is an example of the point

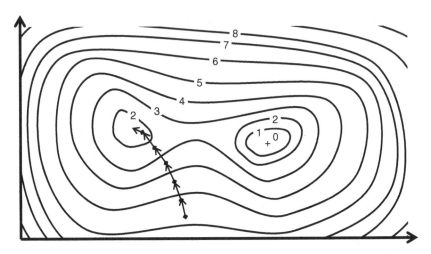

Figure 7 Gradient descent in a two-dimensional parameter space, wherein the
contour lines illustrate different levels of the cost function. From the given
starting point, the algorithm finds a local minimum on the left rather than the
absolute minimum to the right.

difference making to determine the dependencies between input and outcome
variables. Thus, they fit well with variational induction, as I will argue in more
detail in Section 4.3.

4.2.4 Association Rules

As an example of unsupervised learning, so-called association rule algorithms try to find frequent co-occurrences among different variables. An
early application of this method was to find associations between different
items that are bought in a supermarket, say that people tend to buy cigarettes
and beer at the same time (Ghani & Schierholz 2017, 160–1). A typical
association rule algorithm proceeds by the following two steps: first, find all
combinations of items or features that occur with a specified minimum
frequency; second, generate association rules that express co-occurrences
of items or features within these combinations (2017, 161). Thus, association
rules are especially useful for finding co-occurrence patterns between features. Such co-occurrence patterns or association rules typically have the
form $X_1, \ldots, X_n \rightarrow Y$, where all X and Y designate individual features and
n usually is a small number (2017, 160).

---

discussed in Section 6.2 that machine learning algorithms in general do not implement variational induction in its pure form.

Association rules can be evaluated in terms of various criteria including support S, confidence C, and lift L. Support denotes how often certain features are present in a data set. For example, the support of the feature (cigarettes) in a data set corresponds to the number, ratio, or percentage of instances in the data set with feature (cigarettes), e.g., all those transactions in which cigarettes are bought divided by the total number of transactions in the data set. Confidence is a measure for how often a specific rule $X_1, \ldots, X_n \rightarrow Y$ holds (e.g., that beer is bought if cigarettes are bought). Confidence is evaluated as support of feature $(X_1, \ldots, X_n, Y)$ divided by the support of feature $(X_1, \ldots, X_n)$. Lift then is evaluated as support of $(X_1, \ldots, X_n, Y)$ divided by the product of support of $(X_1, \ldots, X_n)$ and support of (Y). Obviously, lift is a measure for how much the co-occurrence pattern of $(X_1, \ldots, X_n)$ and (Y) deviates from the case, where they are independent of each other. In other words, lift serves as an indicator to what extent certain observed correlations are random or not.

In contrast to decision trees and neural networks, the association rules algorithm was discussed here as an example of an algorithm that does not implement a difference-making or variational rationale, but largely adheres to an enumerative approach. In general, scientific practice has shown such algorithms to be much less powerful than those implementing a variational rationale. The results may be good enough for the decision making of a supermarket chain, but not for establishing scientific results.

4.3 Machine Learning as Variational Induction

4.3.1 Machine Learning as Enumerative Induction

The question of what kind of induction machine learning algorithms implement is rarely discussed in the literature, which is unsurprising given the prevalent anti-inductivist attitudes described in Section 3.1. Moreover, the few scholars who address this issue show substantial differences in their assessments. Given that enumerative induction is still widely held to be the most common type of induction, it is not difficult to find arguments that machine learning makes use of enumerative induction. For example, computer scientist and philosopher of science Gregory Wheeler writes that

> [s]upervised learning problems are thus a form of enumerative induction.
> However, instead of observing a finite number of input-output pairs of
> some kind in order to draw an inference about all pairs of that kind, as
> enumerative induction is conceived to do, a supervised learning problem
> typically observes a finite number of pairs to draw an inference about an
> unobserved but likewise finite set of pairs. There is no pretense to learning
> a universal law through supervised learning, and in fact basic results in the

field suggest that a search for universality is entirely the wrong approach. (2016, §5)

In the following, very different conclusions will be reached. While there may be some rather weak and unreliable algorithms that rely on enumerative induction such as the association rules discussed in the previous Section 4.2.4, I will argue that the most successful algorithms like decision trees or neural networks implement a version of variational induction.

Occasionally, the assumption that machine learning relies on enumerative induction is only implicit in the general epistemological outlook. An example in this respect is the work of Ray Solomonoff, who is a central figure for many computer scientists interested in the epistemological foundations of data science. Solomonoff was one of about twenty attendees of the so-called Dartmouth Summer Research Project on Artificial Intelligence in 1956, which is generally considered the founding event of artificial intelligence as a scientific field.

Solomonoff frames inductive inference as the problem of analyzing how often certain types of events reoccur in given sequences of events and then extending those sequences (1964a, 1964b, 1999, 2008; see also Sterkenburg 2016). That, of course, is the starting point of enumerative induction:

> The problem dealt with will be the extrapolation of a long sequence of symbols – these symbols being drawn from some finite alphabet. More specifically, given a long sequence, represented by T, what is the probability that it will be followed by the sequence represented by a? In the language of Carnap [. . .], we want c(a, T), the degree of confirmation of the hypothesis that a will follow, given the evidence that T has just occurred. [. . .] In all cases being considered, the known sequence of symbols is very long, and contains all of the information that is to be used in the extrapolation. (Solomonoff 1964a, 2)

That Solomonoff's approach implements an enumerative rationale is further corroborated by the reference to Rudolf Carnap, whose student Solomonoff was. Carnap's inductive logic was based on the so-called straight rule of induction, which is a core idea of enumerative approaches (Pietsch 2019, Sect. 2.1). Certainly, with some extra constraints and structure in Solomonoff's sequences of symbols, the variational evidence required for variational induction could be mapped onto these sequences, but this would result in a merely formal and unintuitive way of looking at variational induction.

In summary, the way in which Solomonoff frames the epistemological problem of learning about the world does not lend itself easily to discussing variational induction. Generally speaking, it is important to uncover the various underlying assumptions in epistemological frameworks, since these assumptions often predetermine which methodological issues are raised and whether certain methodological problems can be fruitfully discussed or not.

4.3.2 Machine Learning as Eliminative Induction

Another widely held view is that machine learning constitutes a form of eliminative induction, in the sense that various hypotheses are formulated based on large amounts of data, which hypotheses can then eventually be deductively tested and thus eliminated. For example, Emanuele Ratti delineates a general scheme for inductive inferences in machine learning in terms of the following three phases: "1. formulation of an initial set of hypotheses; 2. elimination of false (or less probable) hypotheses; 3. test (validation) of hypotheses not eliminated in phase 2" (2015, 200).

Given that many see in big data approaches novel possibilities for hypothesis generation, as discussed in Section 3.1, it is not surprising that machine learning is frequently interpreted in terms of eliminative induction. Donald Gillies, in a remarkable and insightful early analysis of machine learning from a philosophy of science perspective, has argued that machine learning implements a Baconian approach to induction, which Gillies like many other modern authors construes broadly as a form of eliminative induction. According to Gillies, Bacon suggested a procedure "of generating a number of possibilities from the data and then eliminating some of them by 'exclusion and rejection'" (1996, 67). The various possibilities are in principle generated inductively from the data, while as yet there is "no scheme for classifying [. . .] inductive rules of inference, and showing how they interrelate" (1996, 104). Gillies discusses two main examples of algorithms including ID3, which is a decision tree algorithm. He emphasizes three features as crucial for machine learning: 1. the existence of inductive rules of inference; 2. the role of background knowledge as well as data in these rules; and 3. the role of testing and falsification in the process of iterating a basic inductive rule of inference to produce the final result (1996, 19). Importantly, Gillies notes that both Bacon's approach and modern machine learning algorithms emphasize the mechanical nature of induction, which at least in principle should be amenable to automation (1996, 2).

An example from the computer science literature of an approach broadly situated in the tradition of eliminative induction is the epistemological framework of Vladimir Vapnik, who is one of the inventors of the so-called support vector machines, which are a major class of machine learning algorithms. In his widely cited book *The Nature of Statistical Learning Theory*, Vapnik depicts his general model of learning from examples as having the following three components:

(i) A generator (G) of random vectors $x \in R^n$, drawn independently from a fixed but unknown probability distribution function $F(x)$.

(ii) A supervisor (S) who returns an output value y to every input vector x, according to a conditional distribution function $F(y|x)$, also fixed but unknown.

(iii) A learning machine (LM) capable of implementing a set of functions $f(x,\alpha)$, $\alpha \in A$, where A is a set of parameters.

Given these three elements, the problem of learning from examples is that of choosing from the given set of functions $f(x,\alpha)$ the function that best approximates the supervisor's response. The selection of the best function is usually based on a training set of random, independent, and identically distributed observations (y, x) drawn according to $F(y|x) = F(x) F(y|x)$. Thus, generator and supervisor can be thought of as providing the data, wherein the training set constitutes a representative sample of those data (Vapnik 2000, 17; see also 1999, 988).

A risk functional is then introduced mapping the functions $f(x,\alpha)$ on a risk variable $R(\alpha)$, which is to be minimized with respect to the given training set. The risk variable measures the discrepancy between the supervisor's response to a given input and the response $f(x,\alpha)$ provided by the learning machine. Typically, the best function is determined by those parameters α, for which the discrepancy is the smallest (1999, 988; 2000, 18).

In principle, Vapnik's approach is general enough to provide an underlying framework for all the machine learning algorithms discussed in Section 4.2. Still, his framing of the learning problem includes central characteristics of eliminative induction, which make it more difficult or unintuitive to subsume methods of variational induction under this framework.[28] In particular, a hypothesis space $f(x,\alpha)$ is presupposed in Vapnik's approach and then one or several best hypotheses within this space are determined. As we have seen, the notion of a hypothesis space (i.e., a set of hypotheses, presumably including the correct hypothesis or at least one that accurately reproduces the data) is central to eliminative induction.

Neither for decision trees nor for neural networks is it necessary or even natural to introduce a hypothesis space. To the contrary, we have seen that a crucial advantage of neural networks is that they are able to approximate almost any type of function, i.e., that they are not restricted by some narrow hypothesis space. The notion of a hypothesis space seems to fit more closely with traditional statistical

---

[28] Harman and Kulkarni (2007) believe that Vapnik's approach constitutes a type of enumerative induction, presumably because positive instances in the training set somehow confirm the 'best' hypothesis.

methods such as linear regression, in which a specific class of functions is predetermined from the outset, for example, linear functions. The notion also fits quite well with Vapnik's support vector machines, since the hyperplanes separating different classes of instances in this algorithm can be thought of as different hypotheses among which an optimum in terms of classification is determined. This may serve as further corroboration of the general point that there is coherence between the fundamental epistemological outlook and the specific methods proposed by individual scholars.

There are other aspects of Vapnik's approach, which fit better with variational induction than with eliminative induction. In particular, the use of a risk or loss function allows for incremental learning, as is typical for variational induction. By contrast, eliminative induction emphasizes discontinuities in the learning process, when falsified hypotheses are discarded and new hypotheses are proposed. As in variational induction, falsification of hypotheses does not play a central role in Vapnik's approach. Rather, the 'best' hypothesis is chosen based on a risk variable indicating how well the hypothesis fits a training data set.

Again, this is not to claim that it is impossible to frame variational methods in terms of Vapnik's approach. What I want to emphasize is that implicit epistemological assumptions always shape the methodological outlook and make it easier or more difficult to address certain methodological problems. If machine learning indeed implements variational induction, as I will now argue, then one should be wary of underlying epistemological frameworks with elements from enumerative or eliminative induction.

4.3.3 Machine Learning as Variational Induction

A central thesis of this Element is that the most successful types of machine learning algorithms all implement variational induction, as I will now argue for decision trees and for neural networks. This basic insight then allows systematically addressing a range of other epistemological issues in subsequent sections including the role of causation for machine learning or how much theoretical background knowledge is required.

Decision trees, for instance, do not start with a set of hypotheses that is subsequently tested as would be required by eliminative induction. Instead, they start with the evidence in terms of large data sets. Arguably, this data is typically provided not in the format required for enumerative induction, i.e., in terms of the repetition of similar instances. Rather, the data usually tracks the precise impact that changes in circumstances have on the phenomenon of interest – which is the kind of data useful for variational induction.

Accordingly, decision trees constructively take into account not only positive instances, where the phenomenon of interest is present, but also negative instances, where it is absent. In Section 4.1.3, I had identified this aspect as another important feature distinguishing variational from enumerative induction.[29] Specifically, when the different branches of a tree are constructed based on information gain, the classification of *all* instances improves, including negative ones. Of course, negative and positive instances are required in variational induction, because only the comparison between them allows determining difference makers among the predictor variables.

Relatedly, decision tree algorithms establish the relevance of predictor variables in terms of the respective information gain, i.e., to what extent these variables make a difference for the accuracy of the classification. Only variables that are relevant in this sense figure in the resulting tree model. Furthermore, as shown at the end of Section 4.2.2, decision tree models can be translated into the type of Boolean expressions that generally result from variational induction.

Finally, homogeneity is ensured not by the algorithm itself, but by adequately framing the problem and preparing the data. Most importantly, researchers intuitively try to include in the set of examined variables all those that they deem potentially relevant for the phenomenon of interest.

All these characteristics, the focus on difference making, the determination of relevance or irrelevance of the predictor variables, the Boolean expressions implied by decision tree models, and the implementation of a homogeneity condition, are typical for variational induction but are absent in enumerative and in eliminative induction.

Similarly, neural networks exhibit these main characteristics of variational induction. They also start with the mentioned variational type of evidence. In particular, they constructively take into account negative instances since the loss function discussed in Section 4.2.3 sums over all instances no matter what the value of the target value is. As in the case of decision trees, homogeneity is again ensured by the framing of the problem and the preparation of the data set.

Neural networks also model a distinction between relevant and irrelevant circumstances or predictor variables. Irrelevant circumstances essentially are those that are given weight zero or sometimes approximately zero, while relevant circumstances are given a nonzero weight.

---

[29] Gillies has rightly emphasized the use of positive and negative instances as a distinctive feature of both Bacon's inductive method and of machine learning approaches: "Although [the use of positive and negative instances of a concept in machine learning] may seem an obvious procedure, it is in fact of considerable importance, and was first suggested by Bacon" (1996, 33).

Furthermore, neural network models as discussed in Section 4.2.3 can be translated into the kind of Boolean expressions that result from variational induction. In the case of neural nets, these expressions typically are highly complex, taking into account a large number of predictor variables. The translation is straightforward, if the input vector **x** comprises only binary variables and if the number of possible values of the outcome variable $f(\mathbf{x})$ is finite. If the input vector comprises continuous variables, the method of difference in its basic form is insufficient and recourse must be taken to the method of concomitant variation, a topic that I cannot address here due to lack of space.

An interesting feature of deep learning, which has significantly contributed to the enormous recent success of these algorithms, consists in the hidden layers that are interposed between the input layer and the output layer (cp. Fig. 6). The role of these hidden layers can also be explained in terms of variational induction, in that by means of the hidden layers a hierarchy of features is constructed, which greatly aids in the task of classification. Remember as an example the tree-structured taxonomy, which was used for predicting malignancy of skin cancer (cp. Sections 1 and 3.3). Of course, such hierarchies of features are ubiquitous in human perception and cognition as well. They can be easily reconstructed in terms of the Boolean combination of circumstances, with which causes of phenomena are formulated in variational induction.

Note that the discussed algorithms do not implement variational induction in an ideal way. Instead they have many pragmatic features, which are often adapted to specific contexts of application. Therefore, it is not feasible to derive any of these machine learning algorithms directly from the basic logic of variational induction. Instead, variational induction can provide a framework to determine under what conditions different machine learning algorithms will succeed or not (see Section 6.2).

Furthermore, as already indicated, not all machine learning algorithms implement variational induction. For example, the association rules presented in Section 4.2.4 clearly rely on an enumerative rationale. After all, when such association rules are generated, negative instances are generally not taken into account constructively. Also, the relevance and irrelevance of predictor variables is not systematically studied.

For the epistemological outlook of this Element, it is not problematic that some machine learning algorithms implement an enumerative or eliminative rationale rather than a variational rationale. Importantly, recent practice in machine learning essentially shows that those algorithms implementing variational induction such as decision trees or neural networks are in general much more successful than those relying on eliminative induction or on enumerative induction.

5 Correlation and Causation

The main conclusion of Section 4 that machine learning algorithms implement variational induction allows one to systematically address the question what role causation plays for big data approaches. Because variational induction identifies causal relationships between circumstances and the phenomenon of interest, it follows immediately that, contrary to popular opinion, causality is central for big data approaches relying on machine learning.

5.1 The Causal Nature of Data Science

5.1.1 Anti-Causal Traditions in Science

One epistemological mantra of big data and data science is that allegedly 'correlation replaces causation.' Not only do big data advocates argue for this thesis, but it is also readily taken up by big data critics in order to show the allegedly unscientific nature of big data approaches.

In an infamous but hugely influential article, WIRED editor-in-chief Chris Anderson has claimed that in the age of big data, "[c]orrelation supersedes causation, and science can advance even without coherent models, unified theories, or really any mechanistic explanation at all" (2008). Another quote to the same effect is from Victor Mayer-Schönberger and Kenneth Cukier's widely read account *Big Data: A Revolution That Will Transform How We Live, Work, and Think*, which was a New York Times Best Seller. According to one of their central theses, big data implies

> a move away from the age-old search for causality. As humans we have been conditioned to look for causes, even though searching for causality is often difficult and may lead us down the wrong paths. In a big-data world, by contrast, we won't have to be fixated on causality; instead we can discover patterns and correlations in the data that offer us novel and invaluable insights. The correlations may not tell us precisely why something is happening, but they alert us that it is happening. (2013, 14)

Apparently, these authors are unaware of the fact that science has moved away from causation multiple times in history and that the kind of causal skepticism that they associate with big data is far from novel and by contrast has been commonplace in the twentieth century. Influential philosophers and epistemologists like Ernst Mach (e.g., 1905, 1923) or Bertrand Russell (e.g., 1913) all have argued that causation constitutes a primitive and outdated mode of thinking. Such views are still fashionable today; for example, the Pittsburgh philosopher of science John Norton has defended a view of "causation as folk science" (2007). In the twentieth century, causal skepticism was also widespread among physicists, with prominent figures like Werner Heisenberg claiming that quantum mechanics empirically disproves the law of causality (e.g., 1931).

Maybe most importantly for the issues discussed in this Element, causal skepticism is deeply rooted in the origins of modern statistics. Karl Pearson, who according to the *Encyclopedia Britannica* was "the leading founder of the modern field of statistics,"[30] is the statistical thinker who most clearly embodies this skepticism, famously stating: "Beyond such discarded fundamentals as 'matter' and 'force' lies still another fetish amidst the inscrutable arcana of modern science, namely, the category of cause and effect" (1911, iv). Pearson, of course, rejected causation in favor of correlation, which he considered the more general and encompassing concept:

> Routine in perceptions is a relative term; the idea of causation is extracted by conceptual processes from phenomena, it is neither a logical necessity, nor an actual experience. We can merely classify things as like; we cannot reproduce sameness, but we can only measure how relatively like follows relatively like. The wider view of the universe sees all phenomena as correlated, but not causally related. (1911, 177)

Thus, Pearson wants to replace causality with the more general concept of correlation, i.e., some kind of statistical association, because completely identical causes as well as completely identical effects allegedly do not occur in the world.

5.1.2 Causation as Basis for Manipulation and Prediction

Against this anti-causal tradition, there is a simple and straightforward positive argument that can be made in favor of a crucial role for causation in the sciences. This argument comes down to the question of what the overall function of causal knowledge is. Why is it important to identify a relationship as causal rather than as a mere correlation? Many, also in the current debate on the epistemology of data science, believe that knowing the causal story is largely equivalent with being able to *explain* a phenomenon. Allegedly, without causal knowledge, one can merely describe *how* things are, but cannot explain *why* they are as they are. Mayer-Schönberger and Cukier, in the above quote, clearly rely on such a link between causation and explanation.

However, such views are mistaken about the primary function of causal knowledge. While causal knowledge sometimes contributes to theoretical explanations, causal explanations per se, which explain by stating the causal factors responsible for a phenomenon, often constitute only very poor and

---

[30] www.britannica.com/biography/Karl-Pearson

superficial explanations.[31] For example, decision trees sometimes provide causal explanations by determining which circumstances are relevant for the phenomenon of interest. While decision tree models thus include some information on why a certain phenomenon occurs, they typically do not provide explanations that would be satisfactory for scientists or even lay persons. In particular, they are usually ignorant about the mechanisms linking specific circumstances with the phenomenon of interest.

Thus, causality does not primarily aim at explanation, at answering why-questions. Instead, the distinction, which is most notably established by causation, is that between empirical relationships that allow changing the phenomena in desired ways and those relationships that do not. The philosopher of science Nancy Cartwright has stated this crucial insight in a particularly clear way: "I claim causal laws cannot be done away with, for they are needed to ground the distinction between effective strategies and ineffective ones" (1979, 420). Causal laws describe difference-making relationships between variables in the sense that if a certain variable is changed, this will have an impact on another variable. Pushing the light switch reliably turns on the light, because a causal relationship exists between these variables. By contrast, increasing the stork population does not change human birth rate, because there is no causal relationship, even though a correlation between the respective variables may exist (cf. Höfer et al. 2004). Above all, causation is a guide on how to effectively intervene in the world.

Data science, of course, is fundamentally interested in difference-making relationships, i.e., in how to effectively intervene in phenomena. For example, algorithms are designed to determine the best medicine to cure a certain cancer, to make someone click on a link, buy a certain product, or vote for a certain person. For such applications, correlations are by definition never enough. Rather, these algorithms must at least implicitly establish causal relationships. Such causal knowledge mostly will not be explanatory, but will yield at best poor answers to why-questions – mirroring the asymmetry between predictive power on the one hand and lack of explanations on the other hand that is characteristic for many big data approaches.[32]

---

[31] The distinction between causal and theoretical explanation has been made for example by Nancy Cartwright, who writes: "there are two quite different kinds of things we do when we explain a phenomenon [...] First, we describe its causes. Second, we fit the phenomenon into a theoretical frame" (1983, 16).

[32] Sullivan (2019) argues based on an analysis of deep neural networks that the complexity or the black box nature of big data models does not limit how much understanding such models provide. According to her account, link uncertainty primarily prohibits understanding, i.e., the lack of scientific and empirical evidence supporting the link that connects a model to the target phenomenon.

Causal knowledge turns out to be indispensable not only for effective intervention but also for reliable prediction. In the absence of a causal connection between different variables, including the absence of an indirect connection via common causes, any correlation between those variables, no matter how strong, cannot establish reliable prediction. After all, any such correlation must be purely accidental, and there is no reason to believe that it will continue to hold in the future. Of course, this argument presupposes that there is also no conceptual or definitional connection between the respective variables.

For example, one can predict the weather by looking at a barometer reading, but one cannot change the weather by intervening with the barometer reading. The prediction is reliable, because there is a common cause in terms of barometric pressure. Data science algorithms frequently single out proxies of the actual causes in the case of imperfect data or if the algorithms do not fully implement variational induction. While such relationships do not constitute direct causal links, they nevertheless constitute indirect causal links via common causes; thus, they are also fundamentally of causal nature. If an algorithm aims to predict whether someone will commit a crime, pay back a credit, or is authorized based on her/his face or fingerprint, causation is indispensable as well.

In summary, both central functions of big data approaches, prediction and intervention, require some access to causal knowledge. Intervention requires knowledge of a direct causal connection between the intervening variable and the phenomenon that is intervened upon, while for prediction an indirect causal relationship via common causes is sufficient. This shows that the ubiquitous claim in the big data literature that with large enough data sets correlations replace causation is mistaken because big data approaches aiming at intervention or prediction by definition have to rely on causal knowledge.

5.1.3 Causal Inference and Variational Induction

A further argument against the claim that in big data approaches correlation replaces causation starts from the close link that exists between induction and causation. Indeed, induction, if properly carried out, leads to causal laws. This link between induction and causation is clearly present in Mill's methods of difference and of agreement as well as in the alternative framework of variational induction presented in Section 4.1.4. All these methods explicitly refer to causal relationships. In fact, variational induction based on difference making is particularly well suited to establish this conceptual link between induction and causation. After all, a plausible argument can be made from the difference-making character of variational induction to the interventionist and manipulationist character of causation as delineated above. The difference-making circumstances identified by variational induction are exactly the circumstances that need to be manipulated or intervened upon in order to change a phenomenon.

One might object that some machine learning algorithms aim at classification, while classification is generally considered to be definitional rather than causal. This may hold in particular for some of the clustering algorithms used in machine learning, which identify clusters of instances in feature space. However, when machine learning algorithms aim at predicting some target variable, as is the case for all algorithms discussed in Section 4.2, such predictions are empirical in nature. Based on the argument of Section 5.1.2, empirical predictions have to be causally supported, even if they can also be understood in terms of conceptual relations or definitions. Adequately classifying instances in order to yield reliable predictions is a fundamentally causal endeavour.

According to a related objection, difference making can also be used in the analysis of conceptual relationships and thus might not necessarily establish the causal nature of a relationship. It is indeed a virtue of the variational framework that with some minor modifications it can be used for analyzing definitional and conceptual relationships as well. This is important because sometimes it depends on the viewpoint of the observer whether a relationship is interpreted as empirical or as definitional. With respect to the algorithms discussed in Section 4.2, as already stated, these all analyze empirical relationships supporting predictions and manipulations; such relationships are causal, as shown in Section 5.1.2.

5.2 Arguments against the Correlational Method

Various arguments have been put forward in recent years to formally prove that 'pure' big data approaches are allegedly not feasible. Particularly influential arguments in this respect have been formulated by mathematicians and computer scientists Cristian Calude and Giuseppe Longo (2017), who argue based on theorems from statistical physics that correlations and regularities in big data sets are unreliable, but that scientific meaning needs to be ascribed to those correlations, e.g., in terms of causation. Related arguments have been put forth by philosopher of science Hykel Hosni and physicist Angelo Vulpiani (2018a, 2018b).

As I will argue, such general arguments against the feasibility of big data approaches are often based on the, as we have seen, mistaken assumption that big data approaches merely identify correlations. Relatedly, they presuppose a perspective of enumerative induction and fail to recognize that many big data approaches rely on variational induction. Indeed, Calude and Longo explicitly frame their argument as being aimed at the so-called correlational method:

the correlational method is a scientific praxis with roots going far back in human history. It is also clear that learning how to analyse extremely large data sets correctly and efficiently will play a critical role in the science of the future, and even today. Rather, our aim is to document the danger of allowing the search of correlations in big data to subsume and replace the scientific approach. (2017, 600)

But, as was shown, framing contemporary big data approaches in terms of a purely correlational method is misguided. By relying on variational induction, big data approaches are to some extent able to distinguish causation from correlation.

5.2.1 The Argument Based on Poincaré Recurrence Time

One argument that big data analysis is useless without at the same time providing sufficient theoretical underpinning is based on the so-called ergodic theory of statistical physics (Calude & Longo 2017, Sect. 5; Hosni & Vulpiani 2018a, Sect. 3; 2018b, Sect. 4). For example, Hosni and Vulpiani frame big data approaches as essentially "interested in forecasts such that future states of a system are predicted solely on the basis of known past states" (2018a, 564). In their view, such approaches look for a past state sufficiently similar to a given present state and assume that the system will evolve from the present state in a similar way as it has evolved from the past state. They coin this 'analog prediction.'

The argument based on recurrence times essentially goes as follows. According to Henri Poincaré's recurrence theorem, bounded dynamical systems under certain conditions – in particular, the dynamics need to be volume preserving in phase space[33] – will return arbitrarily close to the initial state after a sufficiently long, but finite time (see Fig. 8). In principle, this seems good news for the big data advocate. In order to predict the further dynamical evolution from a given state based on data, one just has to look up in the respective data set how the system has evolved in the past from that state. Poincaré's recurrence theorem guarantees that such data may be available at least in principle because the system has been arbitrarily close to the given state in the finite past.

The bad news is that even for relatively simple systems the recurrence times are usually forbiddingly large. They quickly become larger than the age of the universe and thus it appears impossible to have sufficient data for the above procedure to work. This insight is mainly based on Kac's Lemma, according to which recurrence times are inversely proportional to the volume in phase space,

---

[33] The phase space is an abstract space, whose dimensions are determined by the degrees of freedom of a system.

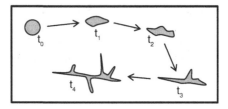

Figure 8 Dynamical evolution over time t_0 to t_4 of a given volume in a two-dimensional phase space bounded by the black box. The dynamics is volume preserving, i.e., the grey regions all have the same area. Under these premises, it is intuitively plausible that points in phase space will eventually return arbitrarily close to their initial states at t_0, as is required by Poincaré's recurrence theorem.

to which an initial condition shall return (Calude & Longo 2017, Sect. 4). In particular, recurrence times scale exponentially with the dimensions of phase space, i.e., essentially the number of variables that are considered (for example, the number of circumstances that are used to predict).

It thus seems implausible to have sufficient data to predict the dynamical evolution just based on data and without any knowledge whatsoever of the underlying dynamics, which supposedly must be derived from the theoretical context. Hosni and Vulpiani conclude that "[a]ll extreme inductivist approaches will have to come to terms with this fact" (2018a, 565). They claim that in the few cases in which analog prediction has worked, the problem necessarily has very low dimension, as, for example, for the prediction of the tides (2018a, 566).

While this argument may have some initial plausibility, the big data success stories from Section 1 already indicate that something is wrong. Contra Hosni and Vulpiani, big data approaches have been successful not only for low-dimensional problems, but also for extremely high-dimensional problems, as in the case of the prediction of skin cancer from images. In this example, a large number of pixels of these images were taken into account to predict, wherein the number of pixels determines the order of magnitude of the dimensionality of the problem.

A central problem with the above argument based on Poincaré's recurrence theorem and Kac's Lemma is that it presupposes a mistaken picture of how machine learning algorithms reason inductively from data. For example, Hosni and Vulpiani claim that "in absence of a theory, a purely inductive modelling methodology can only be based on times series and the method of the analogs, with the already discussed difficulties" (2018a, 568). However, this is clearly an enumerative picture that completely ignores the alternative of variational induction.

Variational induction is capable of determining the relevance or irrelevance of circumstances for specific predictions. If certain circumstances are irrelevant for a specific prediction, then these can be neglected. Thus, any estimate of the amount of data required for predictions based on Kac's Lemma is misguided, since this Lemma takes into account the phase space in its full dimensionality, mistakenly considering *all* circumstances to be equally relevant. Moreover, variational induction can infer causal laws determining the dynamics of a system, which thus does not have to be hypothesized or derived from general theory.

5.2.2 The Curse of Dimensionality

A related argument is based on the so-called 'curse of dimensionality,' an expression coined by Richard Bellman decades ago (1961). According to this concept, the volume of the phase space increases so fast with an increasing number of dimensions that the data in the phase space almost inevitably become sparse even for a relatively small-dimensional feature space. Thus, the number of data required to make predictions becomes forbiddingly large already for a phase space with a relatively small number of dimensions. For example, 10 data points may be sufficient to cover a one-dimensional interval between zero and one. To cover a ten-dimensional feature space, having values between zero and one for each dimension with roughly the same density, 10^{10} (i.e., 10 billion) data points would be needed (see Fig. 9). A ten-dimensional feature space, of course, is still very small in view that many big data applications take hundreds, thousands, or even more features into account.

While sparsity of data often constitutes a problem and for many applications sufficient data are lacking, the curse of dimensionality is not inevitable even in high-dimensional feature spaces for many of the same reasons discussed before. Various approaches for dimensionality reduction are known from the statistics literature, such as principal component analysis, which transforms variables in a way that dependencies are better visible. Similarly, variational induction can, under certain conditions, determine the irrelevance of specific features for given predictive tasks. In this manner, the dimensionality may be reduced for certain predictions such that the data, which may have been sparse in the full feature space, becomes dense enough in the reduced feature space.

A traditional answer to the curse of dimensionality has been to reduce the number of dimensions *overall* to a few that are particularly influential and thus to model the whole phenomenon based on these dimensions. The approach of variational induction and of machine learning algorithms that implement variational induction is different in that, in spite of Bellman's curse, models are constructed that overall take

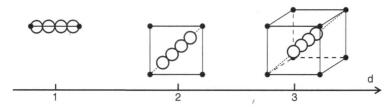

Figure 9 The vicinities of four points cover less and less of the unit volume, when the number *d* of dimensions increases.

into account a vast number of features. In a way, the dimension reduction takes place only locally and in different ways in different regions of the phase space. Decision trees and forests are good examples of this kind of approach, as has already been noted in a classic paper by the statistician Leo Breiman (2001, 208), who himself pioneered work on decision trees and in particular suggested so-called bagging methods, which combine several individual classification or regression tree models into meta-models, which have aptly been termed random forests.

5.2.3 The Argument Based on the Ubiquity of Correlations

A further line of argument, which is also put forward by Calude and Longo, examines to what extent large data sets necessarily comprise spurious correlations, i.e., correlations that emerge randomly and are thus meaningless for prediction. For illustration, they look at strings of numbers, e.g., binary strings, where each element has either value 0 or value 1. A correlation in such a string could, for example, consist in a sequence of *k* elements that have the same value and the same distance to neighboring elements of the same sequence. Such correlations are called arithmetic progressions of length *k*. For example, the binary string 011001101 contains an arithmetic progression of length 3, namely value 1 at positions 3, 6, and 9 of the string. As an example, a value of 1 for a given individual could mean that that individual has been inoculated with smallpox and is resistant to infection, while a value of 0 could mean that the individual is not resistant even though she has been inoculated.

Dutch mathematician Bartel van der Waerden has proven a theorem that for any positive integer *k* there exists a positive integer *g* such that every string of length more than *g* contains an arithmetic progression with *k* occurrences of the same digit. Ramsey's theory generalizes these results for arbitrary correlation functions over *n*-ary relations, i.e., relations that correlate *n* elements of a database (Calude & Longo 2017, Sect. 6).

No one should be surprised that in a large enough database purely random correlations will emerge at least as long as the number of features does not grow in a similar manner. More dubious is Calude and Longo's conclusion that with large enough datasets "most correlations are spurious" (abstract) and thus any true and meaningful relationships will be drowned out by spurious correlations. As has been repeatedly stressed in this Element, for the analysis of epistemological claims in data science the framing of a problem is crucial. The depiction of a data set in terms of a string of symbols, while standard in algorithmic complexity theory, does not lend itself easily to an analysis in terms of variational induction (see the discussion of the work of Solomonoff in Section 4.3.1). Notably, the data for variational induction does not come in a sequence of symbols, but rather in terms of unordered pairs of values (x,y), where x designates all circumstances, features, or predictor variables and y designates the response or target variables. It is not straightforward to map such pairs on the sequences discussed by Calude and Longo.

More importantly, it is not the case that with an increasing amount of the right kind of data, any true relationship will be drowned out by spurious correlations. Rather, if there is sufficient data and if it is correctly prepared,[34] variational induction will single out the 'true' relationships. As a somewhat artificial example, consider a data set that links n circumstances with one variable that is to be predicted. For the sake of simplicity, assume that each of these circumstances and the response variable are all binary variables (i.e., they can take on only two values). Furthermore, the premises required for successful variational induction as laid out in Section 6.2 shall be fulfilled. In particular, homogeneity and causal determination shall hold, i.e., the actual cause of the response variable with respect to the considered context shall be expressible in terms of the given predictor variables. Under such circumstances, at most 2^n data points will be required to determine the true causal relationships by means of variational induction. This is because 2^n is the number of possible combinations of values for n binary predictor variables, thus covering all possible variations of the predictor variables (see the discussion of N=all in Section 2.3). Any further data points will be redundant, and, crucially, any arbitrary amount of further data will not prevent the analysis in terms of variational induction, far from drowning the actual causal relationships in a sea of correlations.

Calude and Longo conclude: "Our work confirms the intuition that the bigger the database which one mines for correlations, the higher is the chance to find recurrent regularities and the higher is the risk of committing [causal] fallacies [such as cum hoc ergo propter hoc; post hoc ergo propter hoc]" (2017, 609). While this conclusion may hold with respect to an impoverished enumerative

---

[34] Data preparation is an immensely important step for tackling big data problems (see Section 6.2 for criteria on how to understand 'correctly prepared' in this context).

methodology, it is clearly mistaken with respect to variational induction and machine learning algorithms relying on variational induction, essentially because the regularities identified by variational induction are causally supported.

Calude and Longo insist that "[i] n other words, there will be regularities, but, by construction, most of the time (almost always, in the mathematical sense), these regularities cannot be used to reliably predict and act" (2017, 609). By contrast, as we have seen, variational induction has no problem singling out the correct causal relationships if they are present in the data, no matter how large the data sets are.

6 The Role of Theory

The central lesson of Section 4 that machine learning implements variational induction also allows one to systematically address another central epistemological question with respect to big data approaches, namely whether and how much theory is required for these approaches. To address this issue, essentially two questions have to be tackled: first, what are the theoretical assumptions required for successful variational induction, and second, how well do specific machine learning algorithms implement variational induction?

6.1 The Alleged Indispensability of Theory

At least since Chris Anderson published his article alleging "the end of theory" (2008), philosophers and scientists alike have discussed the role of theory and modeling assumptions in big data approaches. As in the case of Anderson's other theses, this one as well has met strong reactions from a number of epistemologists and philosophers of science, who have pointed out that framing a scientific problem always requires theoretical background assumptions. For example, Hosni and Vulpiani ask whether it is "really possible to do methodologically sound scientific research starting from 'raw data,' without constructing modelling hypotheses and, therefore, without theory? We think not. [...] data science performs at its best when it goes hand in hand with the subtle art of constructing models" (2018b, 122). Similarly, Rob Kitchin remarks "an inductive strategy of identifying patterns within data does not occur in a scientific vacuum and is discursively framed by previous findings, theories, and training; by speculation that is grounded in experience and knowledge" (2014, 5).

However, such broad claims are as unhelpful as Anderson's bold thesis for understanding the role of theory in big data approaches. James Woodward has made this point very clearly in a related context:

> a major problem with discussions of the role of "theory" in data to phenomena reasoning (or, for that matter, in connection with the so-called theoryladenness of observation) is that philosophers fail to make discriminations

among the very different ways that theories can be "involved" in this process. They also fail to distinguish among very different things that go under the heading of "theory." Different "involvements" by different sorts of theories can have quite different epistemological implications for data to phenomena reasoning. We need to keep these distinct. (2011, 177)

Thus, when we examine the theory-ladenness of big data approaches in the following we will find that certain types of modeling assumptions and background theory are required for reliable results, while others can be dispensed with.

6.2 How Much and What Kind of Theory?

6.2.1 The Problem of Variational Induction

A number of arguments have been raised against inductivism. Classic sources for such arguments are the work of Pierre Duhem, in particular his *Aim and Structure of Physical Theory*, as well as, of course, Popper's work, in particular his main epistemological treatise *The Logic of Scientific Discovery*. Such anti-inductivist arguments include, for example, the alleged theory-ladenness of observation. According to this thesis, which is often traced back to Duhem, there are no brute, uninterpreted facts, from which scientific laws or theories could be inductively inferred. Rather, the facts themselves cannot be portrayed without theoretical background assumptions. Think of the observation of an elementary particle at a particle accelerator such as CERN, which can only be established based on substantial theoretical knowledge of the relevant physical theories. But, in the spirit of Woodward's above quote, such categorical statements are unhelpful for epistemological analyses. Instead, I will examine in detail in Section 6.2.2, which kind of theoretical assumptions have to be presupposed and which can be dispensed with in big data approaches.

By far the most influential argument against inductivism is based on the so-called problem of induction. This problem, which dates back at least to the Scottish philosopher David Hume (1748), is at the root of all skepticism concerning inductivist approaches for scientific methodology. It is the main argument brought forward by Popper in his attack on traditional scientific methodology and his defense of hypothetico-deductivism (1935). The problem of induction is intimately related with the issue that thus far there is no generally accepted inductive method. Indeed, many equate induction with enumerative induction, while at the same time emphasizing the serious deficiencies of this method.

The problem of induction is often framed as the problem of justifying induction, wherein David Hume infamously claimed that there is no justification for induction in general. However, a more interesting question is under what additional assumptions specific inductive methods can be justified. We will see now that the answer differs widely with respect to the different types of induction that were delineated in Section 4.1.

Every inductive method has its own problem of induction, which from a pragmatic viewpoint is adequately framed as the question under which premises reliable inferences result. The traditional problem of induction, notably in Hume's influential formulation, refers mostly to enumerative induction, although this is not always made explicit. Usually, enumerative inferences are taken to presuppose some *principle of the uniformity of nature*, which Hume in his *Treatise of Human Nature* states as follows: "that instances of which we have had no experience, must resemble those, of which we have had experience, and that the course of nature continues always uniformly the same" (cited in Vickers 2018, §2). This is indeed impossible to justify on a general level – essentially due to the vicious circularity that nature is only uniform in those aspects for which inductive inferences are valid.

The problem of eliminative induction shall be only briefly addressed. Obvious difficulties arise in that it does not seem feasible to formulate an exhaustive set of hypotheses based on a given data set. For example, the computer scientist Francesco Bergadano writes: "The traditional treatment of the question of valid induction has underestimated the problem of defining a suitable hypothesis space, and concentrated instead on the issue of justifying the inductive leap from observed to future performance once a particular hypothesis is given" (1993, 47).

The problem of variational induction is completely distinct from the problem of eliminative induction as well as from Hume's problem of enumerative induction. In particular, variational induction does not presuppose an ultimately indefensible uniformity of nature, but rather requires the following assumptions that are more realistic if, however, anything but trivial: (i) causal determination that the examined phenomenon is fully determined by its circumstances; (ii) adequacy of the language with which phenomenon and circumstances are described, wherein such language can in principle be developed in terms of variational induction; (iii) homogeneity or constancy of background as discussed in Sections 4.1.3 and 4.1.4; (iv) sufficient data in terms of instances covering all relevant variations of circumstances (Keynes 1921, Ch. 22; von Wright 1951, Ch. V; Baumgartner & Graßhoff 2003, Sect. IX 2.4; Pietsch 2014, Sect. 3.6).

It can be shown that more data generally leads to more exact causal language and a better understanding to what extent homogeneity is fulfilled. Let us briefly turn to causal determination, which arguably constitutes the crucial premise for variational inferences. This premise is required because of indeterministic situations, where the method of difference and the strict method of agreement fail. For example, a certain indeterministic phenomenon may change due to pure chance while at the same time a circumstance varies that is in fact completely irrelevant to the phenomenon. The method of difference would incorrectly identify this circumstance as a cause of the phenomenon.

From a modern perspective, one might be inclined to think that any method that presupposes causal determination cannot be taken seriously given that many sciences, from quantum mechanics to sociology, assume ubiquitous indeterminism. However, variational induction works for weakened versions of causal determination, for example, the variant of indeterminism present in the orthodox Copenhagen interpretation of quantum mechanics. In this case, the circumstances do not causally determine the specific event that is occurring, but determine a probability distribution over related events. By means of a coarse-grained description subsuming probability distributions under coarse-grained variables, causal determination can be restored in principle and, consequently, variational induction remains applicable.

6.2.2 Theory-Ladenness of Big Data Approaches

The problem of theory-ladenness for big data approaches can be divided into two parts. First, one has to examine to what extent a specific machine learning algorithm implements variational induction, i.e., in which ways the algorithm simplifies or approximates 'pure' variational induction as given by the methods discussed in Sections 4.1.3 and 4.1.4. This first part is specific to the individual algorithm and I will not further discuss it here. Second, one has to determine which theoretical assumptions have to be made in order for 'pure' variational induction to yield reliable results. This second part is universal in the sense that it does not depend on the individual algorithm. This second part comes down to the problem of variational induction as discussed above.

Relying on the above analysis of the problem of variational induction, we can now identify the elements of theory that have to be presupposed in big data approaches based on variational induction. In particular: (a) one has to know all variables A that are potentially relevant for a phenomenon C in a given context provided by a background B; (b) the phenomenon of interest C must be determined by these variables A in the context B; (c) one has to assume that for all collected instances and observations the relevant background conditions remain the same, i.e., a stable context B; (d) one has to have good reasons to expect that the variables A and C are formulated in stable causal categories that are adequate for the addressed research question; (e) there must be a sufficient number of instances to cover all potentially relevant configurations determined by combinations of different values of the variables A (cf. Pietsch 2015). These assumptions are theoretical in a broad sense that they are all to some extent justifiable in terms of a wider theoretical context.

Because these aspects all concern the framing of a research question, one could speak of *external theory-ladenness*. By contrast, *internal theory-ladenness*, i.e., prior assumptions about the causal relationships between the considered variables A and C, is largely absent from big data approaches. For example, in decision trees no prior hypotheses are made about causal connections between the considered variables. Thus, the essential difference in comparison with a hypothesis-driven approach is that not much is presupposed about the internal causal structure of the phenomenon itself. Rather, this structure is mapped from the data by variable variation (Pietsch 2015; 2016a).

In an analysis of several case studies mainly from the social and economical sciences, Robert Northcott has pointed out further issues that may not be fully subsumable under the above-listed requirements (2019). In particular, he notes that measurement errors may arise, that systems may be open or chaotic, or that relationships may be reflexive, i.e., to some extent exhibit a circularity between causes and effects (2019, Sect. 7). A further issue concerns whether the data constitute a representative sample adequately reflecting the relative frequencies with which certain features occur in the population of interest (e.g., 2019, Sect. 2). Note, however, that a representative sample in this conventional statistical sense is *not* required for variational induction. Indeed, the notion of a representative sample is closely tied to an enumerative rather than a variational approach. For the latter, it is crucial that the data represents all relevant changes in circumstances. However, the frequencies with which certain instances occur in the data set are irrelevant for variational induction, at least in a deterministic setting.

In any case, Northcott rightly emphasizes that there is no guarantee that the above conditions for big data approaches based on variational induction will be fulfilled for specific problems. Instead, it may well be that big data do not improve predictions for those scientific tasks for which predictions have always been difficult, such as forecasting the future development of the GDP, which is one of Northcott's examples (2019, Sect. 4). Thus, there is no substitute for local and contextual investigations into whether big data approaches are fruitfully applicable in specific research contexts (2019, Sect. 7). As shown in the introduction to this Element, examples exist of successful scientific practice providing proof of concept that big data approaches work at least in principle. However, in accordance with Northcott's study, the jury is still out as to what extent these approaches are fruitfully applicable in various scientific fields.

7 Novel Scientific Methodology?

Again drawing on the central thesis from Section 4 that machine learning implements variational induction, I now examine to what extent big data approaches rely on novel scientific methodology based on a comparison with established experimental practices.

7.1 Exploratory and Theory-Driven Experimentation

In the following, I argue that so-called *exploratory experimentation* exhibits crucial similarities with big data approaches. Exploratory experimentation relies on an inductivist methodology and must be distinguished from *theory-driven experimentation* relying on a hypothetico-deductive methodology. This is arguably the most fundamental distinction with respect to experimental practice in the sciences (Burian 1997, Steinle 1997).

Before the reign of modern hypothetico-deductivism, methodologists working from an inductivist perspective were clearly aware of exploratory functions of experiments.[35] Ernst Mach, for example, dedicates a long chapter of his book *Knowledge and Error* to the analysis of experimental research and concludes: "What we can learn from an experiment resides wholly and solely in the dependence or independence of the elements or conditions of a phenomenon. By arbitrarily varying a certain group of elements or a single one, other elements will vary too or perhaps remain unchanged. The basic method of experiment is the method of variation" (1905, 149). Mach goes on to discuss whether such variation is feasible given that the number of combinations becomes forbiddingly large already with a few independent variables.

Compare Mach's depiction with a recent account of exploratory experimentation by Friedrich Steinle and Uljana Feest, according to whom exploratory experimentation is "a type of experiment that consists in systematic variation of experimental variables with the aim of phenomenologically describing and connecting phenomena in a way that allows a first conceptual structure of a previously poorly structured research field. [...] exploratory experiments typically take place during historical episodes in which no theories or appropriate conceptual frameworks are available and can result in the formation of such frameworks" (Feest & Steinle 2016, 282).

Both Mach as well as Feest and Steinle emphasize a methodology of variable variation as the core feature of exploratory experimentation. Such variable

---

[35] Even in the twentieth century, scientists who were willing to reflect on their own scientific practice emphasized the exploratory nature of many experiments. A concise account of exploratory experimentation can, for example, be found in Richard Feynman's Caltech Commencement Address (1974).

variation, of course, is at the heart of variational induction and, under certain circumstances as discussed in Section 6, allows one to inductively infer causal relationships governing the examined phenomenon.

Laboratory experiments are typical examples for exploratory experimentation – at least when the causal laws and causal concepts governing the examined phenomena are still largely unknown, i.e., when there is no theory guiding the experimental enquiry. Usually, experimenters first try to determine those variables that may have an impact on the phenomenon based on background knowledge and intuitions. Then, they systematically vary those variables – if possible, only one at a time – in order to determine whether they actually have an impact on the phenomenon. There are plenty of examples for exploratory experimentation (e.g., Röntgen's experiments to understand the new kind of radiation that he had accidentally discovered).

A typical statement of theory-driven experimentation can be found in Karl Popper's work: "The theoretician puts certain questions to the experimenter and the latter by his experiments tries to elicit a decisive answer to these questions, and to no others. [...] Theory dominates the experimental work from its initial planning to the finishing touches in the laboratory" (1935, 89–90). Obviously, this role of experiments fits well with hypothetico-deductivism, of which Popper was a main proponent. A typical example of a theory-driven experiment is Sir Arthur Eddington's 1919 expedition to observe a solar eclipse in order to verify some empirical consequences of Einstein's theory of relativity. In a theory-driven experiment, the respective theory is already well established. The experimental setup is not exploratory – there is little variation of potentially relevant variables. Rather, a theory-driven experiment is generally expected to yield a yes-or-no answer: Does reality correspond to the prediction of the theory or not?

7.2 How Novel Are Big Data Approaches

Based on the above analysis of experimental practices, the question of how novel big data approaches are in terms of scientific methodology can be addressed by a comparison with exploratory experimentation. Indeed, big data approaches share many features with exploratory experimentation. They both rely on the same type of evidence examining a phenomenon under changing circumstances. In both cases, the underlying inferential logic is variational induction. As a consequence, big data approaches and exploratory experimentation are theory independent in similar ways. And finally, they both aim to identify causal structure to warrant manipulation and/or prediction.

There are a number of differences as well, but they are of minor importance compared with the shared reliance on variational induction. One difference concerns the nature of the evidence that is used for inferences. In exploratory experimentation the evidence trivially is mainly of experimental nature, while the data relied upon in big data approaches is often observational.[36] However, this difference has no deeper implications, since it does not matter for variational induction, whether the compared instances result from experiments or from observation. Experimental evidence holds only a pragmatic advantage over observational evidence, since homogeneity is more easily ensured in the former case.

A crucial further difference is that big data approaches can handle a lot more variables than one could squeeze into a laboratory setting and are able to deal with a lot more data than could be handled by human experimenters. Thus, while the underlying inductive logic is not novel but has been used for centuries in exploratory experimentation, big data approaches allow to analyze a wide range of phenomena that are not accessible to conventional exploratory experimentation. This makes a huge difference to scientific practice in data-rich special sciences like medicine or the social sciences.

8 Conclusion

In this Element, some of the major topics and debates in the emerging field of an epistemology of data science and machine learning have been introduced. A sometimes surprising plurality of positions and views was revealed both in the sciences and in the philosophy of science.

I have defended an inductivist view of big data approaches. In fact, successful scientific practice in data science and machine learning constitutes the most convincing argument to date for a revival of inductivist epistemology and for revisiting those arguments based on which inductivism had once been rejected. Of course, this is not to say that inductivist big data approaches are useful for answering all scientific questions. Rather, they should be considered as an addition to the existing methodological toolbox of the sciences.

The development of a coherent conceptual framework for inductivist methodology allowed to systematically address some of the central epistemological questions discussed with respect to big data approaches, namely to what extent data can speak for themselves, whether correlation replaces causation in the wake of big data, and whether the end of theory is near. Conceptual analysis provides a rational and systematic way of addressing these issues, such that the answers do not merely come down to differences in opinion or obsolete epistemological prejudices.

---

[36] Regarding the distinction between experiments and observations, most importantly, experiments require systematic manipulation of the phenomenon of interest by an experimenter, while for observations there is no systematic manipulation.

Much remains to be done. In particular, the link between various machine learning methods and variational induction should be examined in much more detail. Also, the many arguments that have been raised in the past against inductivism and variational induction need to be more comprehensively addressed for a thorough defense of inductivist epistemology.

References

Adriaans, P. (2019). Information. In E. N. Zalta, ed., *The Stanford Encyclopedia of Philosophy (Spring 2019 Edition)*, plato.stanford.edu/archives/spr2019/entries/information/.

Ampère, J.-M. (1826/2012). *Mathematical Theory of Electro-Dynamic Phenomena Uniquely Derived from Experiments*, transl. M. D. Godfrey. Paris: A. Hermann, archive.org/details/AmpereTheorieEn.

Anderson, C. (2008). The end of theory: The data deluge makes the scientific method obsolete. *WIRED Magazine*, **16**/07, www.wired.com/science/discoveries/magazine/16–07/pb_theory.

Bacon, F. (1620/1994). *Novum Organum*. Chicago: Open Court.

Baumgartner, M., & Falk, C. (2019). Boolean difference-making: A modern regularity theory of causation. *The British Journal for the Philosophy of Science*, doi.org/10.1093/bjps/axz047.

Baumgartner, M., & Graßhoff, G. (2003). *Kausalität und kausales Schliessen*. Bern: Bern Studies in the History and Philosophy of Science.

Bellman, R. E. (1961). *Adaptive Control Processes: A Guided Tour*. Princeton: Princeton University Press.

Bergadano, F. (1993). Machine learning and the foundations of inductive inference. *Minds and Machines*, **3**, 31–51.

Bird, A. (2010). Eliminative abduction: Examples from medicine. *Studies in History and Philosophy of Science Part A*, **41**(4), 345–52.

Bogen, J., & Woodward, J. (1988). Saving the phenomena. *The Philosophical Review*, **97**(3), 303–52.

boyd, d., & Crawford, K. (2012). Critical questions for big data. Provocations for a cultural, technological, and scholarly phenomenon. *Information, Communication & Society*, **15**(5), 662–79.

Breiman, L. (2001). Statistical modeling: The two cultures. *Statistical Science*, **16**(3), 199–231.

Burian, R. (1997). Exploratory experimentation and the role of histochemical techniques in the work of Jean Brachet, 1938–1952. *History and Philosophy of the Life Sciences*, **19**, 27–45.

Calhoun, C. (2002). *Dictionary of the Social Sciences*. Oxford: Oxford University Press.

Callebaut, W. (2012). Scientific perspectivism: A philosopher of science's response to the challenge of big data biology. *Studies in History and Philosophy of Biological and Biomedical Science*, **43**(1), 69–80.

Calude, C. S., & Longo, G. (2017). The deluge of spurious correlations in big data. *Foundations of Science*, **22**(3), 595–612.

Cartwright, N. (1979). Causal laws and effective strategies. *Noûs*, **13**(4), 419–37.

Cartwright, N. (1983). *How the Laws of Physics Lie*. Oxford: Oxford University Press.

Clark, A. (1996). Philosophical Foundations. In M. A. Boden, ed., *Artificial Intelligence*. San Diego, CA: Academic Press, pp. 1–22.

Colman, A. M. (2015). *Oxford Dictionary of Psychology*. Oxford: Oxford University Press.

Coveney, P. V., Dougherty, E. R., & Highfield, R. R. (2016). Big data needs big theory too. *Philosophical Transactions of the Royal Society A*, **374**, 20160153.

Duhem, P. (1906/1962). *The Aim and Structure of Physical Theory*. New York: Atheneum.

Einstein, A. (1934). On the method of theoretical physics. *Philosophy of Science*, **1**(2), 163–9.

Esteva, A., Kuprel, B., Novoa, R. A., Ko, J., Swetter, S. M., Blau, H. M., & Thrun, S. (2017). Dermatologist-level classification of skin cancer with deep neural networks. *Nature*, **542**, 115–18.

Feest, U., & Steinle, F. (2016). Experiment. In P. Hymphreys, ed., *The Oxford Handbook of Philosophy of Science*. Oxford: Oxford University Press, pp. 274–95.

Feynman, R. (1974). Cargo cult science. *Engineering and Science*, **37**(7), 10–13.

Flach, P. (2012). *Machine Learning: The Art and Science of Algorithms That Make Sense of Data*. Cambridge: Cambridge University Press.

Floridi, L. (2008). Data. In W. A. Darity, ed., *International Encyclopedia of the Social Sciences*. Detroit: Macmillan.

Floridi, L. (2011). *The Philosophy of Information*. Oxford: Oxford University Press.

Floridi, L. (2019). Semantic conceptions of information. In E. N. Zalta, ed., *The Stanford Encyclopedia of Philosophy (Winter 2019 Edition)*, plato.stanford. edu/archives/win2019/entries/information-semantic/.

Foster, I., Ghani, R., Jarmin, R. S., Kreuter, F., & Lane, J. (2017). *Big Data and Social Science*. Boca Raton, FL: CRC Press.

Foster, I., & Heus, P. (2017). Databases. In I. Foster, R. Ghani, R. S. Jarmin, F. Kreuter, & J. Lane, eds., *Big Data and Social Science*. Boca Raton, FL: CRC Press, pp. 93–124.

Frické, M. (2014). Big data and its epistemology. *Journal of the Association for Information Science and Technology*, **66**(4), 651–61.

Ghani, R., & Schierholz, M. (2017). Machine learning. In I. Foster, R. Ghani, R. S. Jarmin, F. Kreuter, & J. Lane, eds., *Big Data and Social Science*. Boca Raton, FL: CRC Press, pp. 147–86.

Gillies, D. (1996). *Artificial Intelligence and Scientific Method*. Oxford: Oxford University Press.

Goodfellow, I., Bengio, Y., & Courville, A. (2016). *Deep Learning*. Cambridge, MA: Massachusetts Institute of Technology Press.

Graßhoff, G., & May, M. (2001). Causal regularities. In W. Spohn, M. Ledwig, & M. Esfeld, eds., *Current Issues in Causation*. Paderborn: Mentis Verlag, pp. 85–114.

Hacking, I. (1992). The self-vindication of the laboratory sciences. In A. Pickering, ed., *Science as Practice and Culture*. Chicago: Chicago University Press, pp. 29–64.

Hambling, D. (2019). The Pentagon has a laser that can identify people from a distance – by their heartbeat. *MIT Technology Review*, www.technologyre view.com/2019/06/27/238884/the-pentagon-has-a-laser-that-can-identify-people-from-a-distanceby-their-heartbeat/.

Harman, G., & Kulkarni, S. (2007). *Reliable Reasoning. Induction and Statistical Learning Theory*. Boston: Massachusetts Institute of Technology Press.

Hastie, T., Tibshirani, R., & Friedman, J. (2001). *The Elements of Statistical Learning*. New York: Springer.

Heisenberg, W. (1931). Kausalgesetz und Quantenmechanik. *Erkenntnis*, **2**, 172–82.

Hempel, C. G. (1966). *Philosophy of Natural Science*. Upper Saddle River, NJ: Prentice Hall.

Höfer, T., Przyrembel, H., & Verleger, S. (2004). New evidence for the theory of the stork. *Paediatric and Perinatal Epidemiology*, **18**(1), 88–92.

Holland, P. W. (1986). Statistics and causal inference. *Journal of the American Statistical Association*, **81**(396), 945–60.

Hosni, H., & Vulpiani, A. (2018a). Forecasting in light of big data. *Philosophy & Technology*, **31**, 557–69.

Hosni, H., & Vulpiani, A. (2018b). Data science and the art of modelling. *Lettera Matematica*, **6**, 121–9.

Hume, D. (1748). *An Enquiry Concerning Human Understanding*. London: A. Millar.

Jelinek, F. (2009). The dawn of statistical ASR and MT. *Computational Linguistics*, **35**(4), 483–94.

Keynes, J. M. (1921). *A Treatise on Probability*. London: Macmillan.

Kitchin, R. (2014). *The Data Revolution*. Los Angeles: Sage.

Knüsel, B., Zumwald, M., Baumberger, C., Hirsch Hadorn, G., Fischer, E., Bresch, D., & Knutti, R. (2019). Applying big data beyond small problems in climate research. *Nature Climate Change*, **9**, 196–202.

Kohavi, R., Tang, D., & Xu, Y. (2020). *Trustworthy Online Controlled Experiments: A Practical Guide to A/B Testing*. Cambridge: Cambridge University Press.

Kuhlmann, M. (2011). Mechanisms in dynamically complex systems. In P. Illari, F. Russo, & J. Williamson, eds., *Causality in the Sciences*. Oxford: Oxford University Press.

Laney, D. (2001). *3D Data Management: Controlling Data Volume, Velocity, and Variety. Research Report.* blogs.gartner.com/doug-laney/files/2012/01/ad949-3D-Data-Management-Controlling-Data-Volume-Velocity-and-Variety.pdf

Lavoisier, A. (1789/1890). *Elements of Chemistry*. Edinburgh: William Creech.

Lazer, D., Kennedy, R., King, G., & Vespignani, A. (2014). The parable of Google Flu: Traps in big data analysis. *Science*, **343**(6167), 1203–5.

LeCun, Y., Bengio, Y., & Hinton, G. (2015). Deep learning. *Nature* **521**, 436–44.

Leonelli, S. (2014). What difference does quantity make? On the epistemology of big data in biology. *Big Data & Society* **1**(1).

Leonelli, S. (2016). *Data-Centric Biology: A Philosophical Study*, Chicago: Chicago University Press.

Leonelli, S. (2019). What distinguishes data from models? *European Journal for Philosophy of Science* **9**, 22.

Luca, M., & Bazerman, M. H. (2020). *Power of Experiments: Decision Making in a Data-Driven World*. Cambridge, MA: Massachusetts Institute of Technology Press.

Lyon, A. (2016). Data. In P. Humphreys, ed., *The Oxford Handbook of Philosophy of Science*. Oxford: Oxford University Press.

Mach, E. (1905/1976). *Knowledge and Error: Sketches on the Psychology of Enquiry*. Dordrecht: D. Reidel.

Mach, E. (1923/1986). *Principles of the Theory of Heat – Historically and Critically Elucidated*, transl. T. J. McCormack. Dordrecht: D. Reidel.

Mackie, J. L. (1967). Mill's methods of induction. In P. Edward, ed., *The Encyclopedia of Philosophy, Vol. 5*. New York: MacMillan, pp. 324–32.

Mackie, J. L. (1980). *The Cement of the Universe*. Oxford: Clarendon Press.

Mayer-Schönberger, V., & Cukier, K. (2013). *Big Data*. London: John Murray.

Mazzocchi, F. (2015). Could big data be the end of theory in science? A few remarks on the epistemology of data-driven science. *EMBO Reports*, **16**(10), 1250–5.

Mill, J. S. (1886). *System of Logic*. London: Longmans, Green & Co.

Minsky, M. L., & Papert, S. A. (1969). *Perceptrons. An Introduction to Computational Geometry*. Cambridge: Massachusetts Institute of Technology Press.

Napoletani, D., Panza, M., & Struppa, D. C. (2011). Toward a philosophy of data analysis. *Foundations of Science*, **16**(1), 1–20.

Ng, A., & Soo, K. (2017). *Numsense! Data Science for the Layman*. Seattle, WA: Amazon.

Northcott, R. (2019). Big data and prediction: Four case studies. *Studies in History and Philosophy of Science A*. doi:10.1016/j.shpsa.2019.09.002

Norton, J. D. (1995). Eliminative induction as a method of discovery: Einstein's discovery of General Relativity. In J. Leplin, ed., *The Creation of Ideas in Physics: Studies for a Methodology of Theory Construction*. Dordrecht: Kluwer Academic Publishers, pp. 29–69.

Norton, J. D. (2005). A little survey of induction. In P. Achinstein, ed., *Scientific Evidence: Philosophical Theories and Applications*. Baltimore: Johns Hopkins University Press, pp. 9–34.

Norton, J. D. (2007). Causation as folk science. *Philosophers' Imprint*, **3**, 4.

Norvig, P. (2009). Natural language corpus data. In T. Segaran & J. Hammerbacher, eds., *Beautiful Data*. Sebastopol, CA: O'Reilly, pp. 219–42.

Panza, M., Napoletani, D., & Struppa, D. (2011). Agnostic science. Towards a philosophy of data analysis. *Foundations of Science*, **16**(1), 1–20.

Pearson, K. (1911). *The Grammar of Science*, 3rd ed., Black.

Pietsch, W. (2014). The structure of causal evidence based on eliminative induction. *Topoi*, **33**(2), 421–35.

Pietsch, W. (2015). Aspects of theory-ladenness in data-intensive science. *Philosophy of Science* **82**(5): 905–16.

Pietsch, W. (2016a). The causal nature of modeling with big data. *Philosophy & Technology*, **29**(2), 137–71.

Pietsch, W. (2016b). A difference-making account of causation, philsci-archive. pitt.edu/11913/.

Pietsch, W. (2017). Causation, probability, and all that: Data science as a novel inductive paradigm. In M. Dehmer & F. Emmert-Streib, eds., *Frontiers in Data Science*. Boca Raton, FL: CRC Press, pp. 329–53.

Pietsch, W. (2019). A causal approach to analogy. *Journal for General Philosophy of Science*, **50**(4), 489–520.

Plantin, J. C., & Russo, F. (2016). D'abord les données, ensuite la méthode? Big data et déterminisme en sciences sociales. *Socio*, **6**, 97–115.

Popper, K. (1935/2002). *The Logic of Scientific Discovery*. London: Routledge Classics.

Ratti, E. (2015). Big data biology: Between eliminative inferences and exploratory experiments. *Philosophy of Science*, **82**(2), 198–218.

Rheinberger, H.-J. (2011). Infra-experimentality: From traces to data, from data to patterning facts. *History of Science*, **49**(3), 337–48.

Rosenblatt, F. (1962). *Principles of Neurodynamics: Perceptrons and the Theory of Brain Mechanisms*, Washington, DC: Spartan Books.

Russell, B. (1913). On the notion of cause. *Proceedings of the Aristotelian Society*, **13**, 1–26.

Russell, S., & Norvig, P. (2009). *Artificial Intelligence*. Upper Saddle River, NJ: Pearson.

Russo, F. (2007). The rationale of variation in methodological and evidential pluralism. *Philosophica*, **77**, 97–124.

Russo, F. (2009). *Causality and Causal Modelling in the Social Sciences. Measuring Variations*, New York: Springer.

Scholl, R. (2013). Causal inference, mechanisms, and the Semmelweis case. *Studies in History and Philosophy of Science Part A*, **44**(1), 66–76.

Schurz, G. (2014). *Philosophy of Science: A Unified Approach*, New York, NY: Routledge.

Solomonoff, R. (1964a). A formal theory of inductive inference, part I. *Information and Control*, **7**(1), 1–22.

Solomonoff, R. (1964b). A formal theory of inductive inference, part II. *Information and Control*, **7**(2), 224–54.

Solomonoff, R. (1999). Two kinds of probabilistic induction. *The Computer Journal*, **42**(4), 256–9.

Solomonoff, R. (2008). Three kinds of probabilistic induction: Universal distributions and convergence theorems. *The Computer Journal*, **51**(5), 566–70.

Steinle, F. (1997). Entering new fields: Exploratory uses of experimentation. *Philosophy of Science* **64**, S65–S74.

Sterkenburg, T. F. (2016). Solomonoff prediction and Occam's Razor. *Philosophy of Science* **83**(4), 459–79.

Sullivan, E. (2019). Understanding from machine learning models. *The British Journal for the Philosophy of Science*, axz035, https://doi.org/10 .1093/bjps/axz035.

Vapnik, V. N. (1999). An overview of statistical learning theory. *IEEE Transactions on Neural Networks*, **10**(5), 988–99.

Vapnik, V. N. (2000). *The Nature of Statistical Learning Theory*, 2nd ed., New York: Springer.

Vickers, J. (2018). The problem of induction. In E. N. Zalta, ed., *The Stanford Encyclopedia of Philosophy (Spring 2018 Edition)*, plato.stanford.edu/arch-ives/spr2018/entries/induction-problem/.

Vo, H., & Silva, C. (2017). Programming with Big Data. In I. Foster, R. Ghani, R. S. Jarmin, F. Kreuter, & J. Lane, eds., *Big Data and Social Science*. Boca Raton, FL: CRC Press, pp. 125–44.

Wan, C., Wang, L., & Phoha, V. (2019). A survey on gait recognition. *ACM Computing Surveys*, **51**(5), 89.

Wheeler, G. (2016). Machine epistemology and big data. In L. McIntyre & A. Rosenberg, eds., *The Routledge Companion to Philosophy of Social Science*. London: Routledge.

Williamson, J. (2004). A dynamic interaction between machine learning and the philosophy of science. *Minds and Machines*, **14**(4), 539–49.

Williamson, J. (2009). The philosophy of science and its relation to machine learning. In M. M. Gaber, ed., *Scientific Data Mining and Knowledge Discovery: Principles and Foundations*. Berlin: Springer, pp. 77–89.

Woodward, J. (2011). Data and phenomena: A restatement and a defense. *Synthese*, **182**, 165–79.

von Wright, G. H. (1951). *A Treatise on Induction and Probability*. New York: Routledge.

Yu, K.-H., Zhang, C., Berry, G. J., Altman, R. B., Ré, C., Rubin, D. L., & Snyder, M. 2016. Predicting non-small cell lung cancer prognosis by fully automated microscopic pathology image features. *Nature Communications*, **7**, 12474.

Zeiler, M. D., & Fergus, R. (2014). Visualizing and understanding convolutional networks. In D. Fleet, T. Pajdla, B. Schiele, & T. Tuytelaars, eds., *Computer Vision – ECCV 2014*. New York, NY: Springer, pp. 818–33.

Cambridge Elements ≡

Philosophy of Science

Jacob Stegenga
University of Cambridge

Jacob Stegenga is a Reader in the Department of History and Philosophy of Science at the University of Cambridge. He has published widely on fundamental topics in reasoning and rationality and philosophical problems in medicine and biology. Prior to joining Cambridge he taught in the United States and Canada, and he received his PhD from the University of California San Diego.

About the Series

This series of Elements in Philosophy of Science provides an extensive overview of the themes, topics and debates which constitute the philosophy of science. Distinguished specialists provide an up-to-date summary of the results of current research on their topics, as well as offering their own take on those topics and drawing original conclusions.

Cambridge Elements ≡

Philosophy of Science

Printed in the United States
By Bookmasters